Simon Gruber

Contrails and Climate Engineering

Process Studies on Natural and Artificial High-Level Clouds and Their Impact on the Radiative Fluxes

Wissenschaftliche Berichte des Instituts für Meteorologie und Klimaforschung des Karlsruher Instituts für Technologie (KIT)
Band 79

Herausgeber: Prof. Dr. Ch. Kottmeier

Institut für Meteorologie und Klimaforschung
am Karlsruher Institut für Technologie (KIT)
Kaiserstr. 12, 76128 Karlsruhe

Eine Übersicht aller bisher in dieser Schriftenreihe erschienenen Bände finden Sie am Ende des Buches.

Contrails and Climate Engineering

Process Studies on Natural and Artificial High-Level Clouds and Their Impact on the Radiative Fluxes

by
Simon Gruber

KIT Scientific Publishing

Dissertation, Karlsruher Institut für Technologie
KIT-Fakultät für Physik

Tag der mündlichen Prüfung: 14. Dezember 2018
Referent: Prof. Dr. Ch. Kottmeier
Korreferent: Prof. Dr. Th. Leisner

Impressum

KIT Scientific Publishing

Karlsruher Institut für Technologie (KIT)
KIT Scientific Publishing
Straße am Forum 2
D-76131 Karlsruhe

KIT Scientific Publishing is a registered trademark
of Karlsruhe Institute of Technology.
Reprint using the book cover is not allowed.

www.ksp.kit.edu

Print on Demand 2019 – Gedruckt auf FSC-zertifiziertem Papier

ISSN 0179-5619
ISBN 978-3-7315-0896-0
DOI 10.5445/KSP/1000090833

Contrails and Climate Engineering - Process Studies on Natural and Artificial High-Level Clouds and Their Impact on the Radiative Fluxes

Zur Erlangung des akademischen Grades eines

DOKTORS DER NATURWISSENSCHAFTEN
(Dr. rer. nat.)

von der KIT-Fakultät für Physik des

Karlsruher Instituts für Technologie (KIT)

genehmigte

DISSERTATION

von

M.Sc. (Meteorologie) Simon Gruber

aus Lahr / Schwarzwald

Tag der mündlichen Prüfung: 14. Dezember 2018

Referent: Prof. Dr. Ch. Kottmeier

Korreferent: Prof. Dr. Th. Leisner

Zusammenfassung

Aktivitäten einer hochentwickelten und global vernetzten Gesellschaft beeinflussen die Atmosphäre unseres Planeten auf vielfältige Art und Weise. In einem komplexen System wie der Erdatmosphäre erzeugt jede Art des Eingriffs vielfältige Rückkopplungsprozesse. Speziell die seit Jahrzehnten ansteigenden anthropogenen Emissionen verändern die Zusammensetzung von atmosphärischen Spurengasen und Aerosolen und führen schlussendlich, durch Veränderung des Strahlungshaushalts, zum Klimawandel.

In verschiedener Hinsicht unterscheiden sich die Emissionen des Luftverkehrs von anderen Quellen, da hier für den Strahlungshaushalt entscheidende Prozesse in unbelasteten Regionen und in großer Höhe angestoßen werden. Exemplarisch werden zwei Aspekte des Themenkomplexes in dieser Arbeit detailliert untersucht.

Im Rahmen der durch den Flugverkehr initiierten atmosphärischen Prozesse spielt die Entstehung von sogenannten Kondensstreifen aus der Abgasfahne von Flugzeugen eine herausragende Rolle, da hierbei eine neue, vollständig künstliche Art hoher Wolken erzeugt wird. Die Bedeutung von Kondensstreifen und durch die Luftfahrt erzeugte Bewölkung hinsichtlich ihres Einflusses auf die atmosphärischen Strahlungsflüsse ist noch immer Gegenstand der Forschung und daher mit Unsicherheiten behaftet.

Im ersten Teil dieser Arbeit kommt das hochauflösende regionale numerische Modell COSMO-ART zum Einsatz, welches auch Anwendung in der operationellen Wettervorhersage findet. Das Modellsystem wird um eine Parametrisierung erweitert, die es erlaubt, sowohl die Entstehung, als auch den Lebenszyklus von Kondensstreifen und Kondensstreifenzirren zu berechnen.

Ein Zweimomentenschema für die Wolkenmikrophysik, erweitert um eine separate Klasse für Kondensstreifeneis, beschreibt diesen Lebenszyklus. Dies ist notwendig, um eine bessere Beschreibung der in Kondensstreifen auftretenden hohen Eiskristallanzahlkonzentrationen zu erreichen. Der zugrundeliegende Satz an Eingabedaten enthält räumlich und zeitlich hochaufgelöste Flugtrajektorien über Zentraleuropa und ist von Echtzeitdaten abgeleitet. Die Parametrisierung stellt damit flugzeugtypabhängige Quellterme für Eismasse und -anzahl in Kondensstreifen bereit.

Um den Einfluss von Kondensstreifen und Kondensstreifenzirren auf die kurzwelligen Strahlungsflüsse an der Erdoberfläche zu untersuchen, wird eine Fallstudie durchgeführt. Diese Arbeit ist die erste Studie, bei der Kondensstreifen und durch den Luftverkehr beeinflusste Wolken mit einem solchen Modellsystem untersucht werden und stellt den ersten Ansatz dar, die Lücke zwischen hochaufgelösten Grobstruktursimulation und Klimastudien zu schließen. Ein besonderer Fokus liegt auf der Demonstration, dass sich die Wettervorhersage hinsichtlich der Berechnung hoher Bewölkung und den resultierenden Strahlungsflüssen verbessert, wenn Kondensstreifen und Kondensstreifenzirren berücksichtigt werden. Dies befähigt das Modell, hohe Wolken zu simulieren, die andernfalls fehlen. Der Effekt der zusätzlichen Bewölkung bewirkt eine Reduzierung sowohl der einfallenden solaren Einstrahlung am Boden, als auch der Energieproduktion mittels Photovoltaik um bis zu 10 %.

Der zweite Teil dieser Arbeit widmet sich der Untersuchung jener mikrophysikalischen Prozesse, welche in der oberen Troposphäre durch künstliche Aerosolpartikel auftreten, die zum Zwecke der Modifikation von Zirren durch kommerzielle Flugzeuge im Winter über der Arktis ausgebracht werden könnten. Aufgrund der Annahme, dass Eiskristalle in Zirren hier hauptsächlich durch homogene Nukleation entstehen, könnten zusätzliche Aerosolpartikel zu erhöhten heterogenen Nukleationsraten führen. Hierbei bilden sich weniger, jedoch größere Eiskristalle, die schneller sedimentieren. Zusätzlich hätten solche veränderten Wolken eine geringere optische Dicke und erlaubten so eine verstärkte langwellige Abstrahlung von Erdoberfläche und Atmosphäre ins All. Großskalig angewandt, könnte somit eine Reduzierung oder zumindest eine Verlangsamung der Erderwärmung erreicht werden.

Im Rahmen einer Fallstudie werden in dieser Arbeit konvektions- und wolkenauflösende Simulationen für einen Bereich der winterlichen Arktis mit dem atmosphärischen Modellsystem ICON-ART durchgeführt. Um realistische Strahlungsflüsse zu erhalten, berücksichtigt eine umfassend erweiterte Parametrisierung der optischen Eigenschaften von Wolken explizit sowohl die Größenverteilung als auch die Form der Hydrometeore. Ein umfangreicher Vergleich mit Satellitendaten und Flugzeugmessungen wird zur Validierung der Simulationen herangezogen.

Vorherige Studien zum Klimaeffekt dieser Methode, die auf vereinfachenden Annahmen fußten und mit gröberer horizontaler Auflösung durchgeführt wurden, führten zu sich teilweise widersprechenden Ergebnissen. Weiterhin ist diese Studie die erste, bei der sowohl ein potentiell realistisches Emissionsszenario, als auch eine prognostische Beschreibung der aerosoldynamischen Prozesse für das auszubringende Material verwendet.

Der primäre Effekt des Impfens von Zirren ist eine Abnahme der Eiskristallanzahlkonzentrationen um bis zu 80 %. Die hierdurch um bis zu 8 W m^{-2} erhöhten Strahlungsflüsse am Oberrand der Atmosphäre erzeugen dabei eine kühlende Wirkung. Zusätzlich wird ein sekundärer Effekt beschrieben. Die durch das ausgebrachte Aerosol gebildeten Eiskristalle führen zu erhöhten Raten des Bereifens von Wolkentropfen in der planetaren Grenzschicht, wodurch sich der Bedeckungsgrad von Mischphasenwolken um bis zu 30 % reduziert. Die resultierende verstärkte langwellige Ausstrahlung am Boden erzeugt einen zusätzlichen kühlenden Effekt. Die Effektivität des Impfens von Zirren ist in einem bestimmten Maße unabhängig von den atmosphärischen Hintergrundbedingungen, skaliert mit der Anzahlkonzentration des ausgebrachten Aerosols und ist am größten für größere Partikel.

Zusätzlich wird eine Anwendung der Methode auf die gesamte Arktis während des Winterhalbjahrs betrachtet, um robustere Erkenntnisse zu erzielen. Hierbei helfen eine längere Zeitspanne und ein größeres Untersuchungsgebiet dabei, verlässlichere Abschätzungen des Einflusses auf Zirren, Mischphasenwolken und die resultierenden Änderungen der Strahlungsflüsse zu erzielen. Ausdünnen von sowohl hoher, als auch tiefer Bewölkung führt zu einem Anstieg der ausgehenden langwelligen Strahlung am Oberrand der Atmosphäre von ungefähr 3 W m^{-2}. Dies ist, konsistent zu früheren Studien, vergleichbar mit dem mittleren, globalen Langzeiteffekt einer Verdopplung der CO_2 Konzentration. Zusätzlich steigt die ausgehende langwellige Strahlung am Boden um durchschnittlich 4 W m^{-2} an.

Die vorliegende Arbeit ist jedoch nicht als eine Befürwortung dieser Methode des Climate Engineering zu verstehen. Vielmehr zeigen die erzielten Erkenntnisse den Bedarf weitergehender Studien an. So kann eine umfassendere Charakterisierung der arktischen Aerosolbeladung, wie auch ein durch Messungen und Laborexperimente vertieftes Verständnis der Eisbildung, welche den Angriffspunkt der Methode ausmacht, dazu beitragen, Unsicherheiten zu reduzieren. Weiterhin sind die mikrophysikalischen Prozesse, welche in Mischphasenwolken ablaufen, noch immer mit Unsicherheiten behaftet und nicht vollständig verstanden.

Abstract

Activities of a highly-developed and hence globally connected society influence the Earth's atmosphere in various ways. In such a complex system, every kind of perturbation triggers many-faceted feedback mechanisms. Especially anthropogenic emissions, increasing since decades, change the composition of atmospheric trace gases and aerosol, ultimately leading to climate change due to a modified global radiation budget.

In several aspects, aircraft emissions differ from other sources, as here, in pristine regions and high altitudes, processes playing a decisive role for the radiation budget of the atmosphere, are directly initiated. In this thesis, two exemplary aspects of this topic are investigated in greater detail.

Within the scope of aviation induced atmospheric processes, the formation of so-called contrails in the exhaust plume of aircraft is of outstanding importance, as a new and entirely artificial type of high-level clouds is generated. The classification of contrails and aircraft induced cloudiness with respect to their impact on the atmospheric radiative fluxes is part of ongoing research and still afflicted with uncertainties. In the first part of this thesis, the high resolution, regional-scale numerical model COSMO-ART that also is used in operational weather forecast, is applied. The model system is extended by a parameterization that allows for both the generation and the life cycle of contrails and contrail cirrus to be calculated. The life cycle of contrails and contrail cirrus is described by a two-moment cloud microphysical scheme that is extended by a separate contrail ice class for a better representation of the high concentration of small ice crystals that occur in contrails. The basic input data set contains the spatially and temporally highly resolved flight trajectories over

Central Europe derived from real time data. The parameterization provides aircraft-dependent source terms for contrail ice mass and number.
A case study is performed to investigate the influence of contrails and contrail cirrus on the shortwave radiative fluxes at the Earth's surface. This is the first time, aviation-induced processes are studied with this kind of model system, hence starting a first attempt closing the gap between high-resolved large-eddy simulations and spatially coarse climate projections. A special focus is put on the demonstration of an improved weather forecast with respect to predicting high-level cloudiness and the resulting radiative fluxes when including contrails and contrail cirrus. Accounting for contrails produced by aircraft enables the model to simulate high clouds that are otherwise missing on this day. The effect of these extra clouds is to reduce the incoming shortwave radiation at the surface as well as the production of photovoltaic power by up to 10 %.
The second part of this thesis is dedicated to an investigation of the microphysical processes occurring when using commercial aircraft for introducing artificial aerosol particles into the hibernal Arctic upper troposphere with the aim of modifying cirrus clouds. Based on the assumption that ice crystals in cirrus clouds here form mainly by homogeneous nucleation, additional aerosol particles could lead to enhanced rates of heterogeneous nucleation, hence creating less but larger and faster sedimenting ice crystals. In addition, cirrus clouds modified in such a way could have a decreased optical depth and therefore would lead to enhanced outgoing longwave radiative fluxes into space. Applied on a larger scale, such a method could aid reducing, or at least decelerating, global warming.
Within a case study, convection- and cloud-resolving simulations of a case study for a limited area of the hibernal Arctic are performed with the atmospheric modeling system ICON-ART. In order to acquire realistic radiative fluxes, the model is extended for a comprehensive parameterization of the cloud optical properties accounting explicitly for the hydrometeors' size distributions and shapes. A thorough comparison with data both

from satellite as well as aircraft measurement is presented to validate the simulations. Former modeling studies investigating the climate effect of this method were performed with simplifying assumptions and coarser resolution, reaching partly contradicting conclusions concerning the method's effectiveness. Besides, this is the first time, both a realistic emission scenario and a prognostic description of aerosol dynamical processes for the seeding material are incorporated for investigating the effects of seeding cirrus clouds. The primary effect of seeding is found to be a reduction of ice crystal number concentrations in cirrus clouds of up to 80 %, leading to increased radiative fluxes at the top of the atmosphere of up to 8 W m^{-2}, and thereby creating a cooling effect. Furthermore, a secondary effect is found, as ice crystals formed from the injected seeding aerosol particles lead to enhanced riming of cloud droplets within the planetary boundary layer, hence effectively reducing the coverage of mixed-phase clouds about 30 %, thus generating additional cooling by enhancing upward longwave radiative fluxes at the surface. The efficacy of seeding cirrus clouds proves to be relatively independent from the atmospheric background conditions, scales with their number concentrations and is highest for large aerosol particles.

These findings are extrapolated to the entire Arctic during wintertime in order to warrant their robustness. Considering both a longer time span and a larger area, it is possible to give a more reliable estimate of the impact on cirrus clouds, mixed-phase clouds and the resulting changes of the radiative fluxes. Thinning of both high- and low-level clouds leads to an increase in outgoing longwave radiation at the top of the atmosphere of about 3 W m^{-2}, which is, consistent to former studies, in the order of the long-term global mean impact of doubled CO_2 concentrations. In addition, a strong average increase of upward longwave radiation at the surface of about 4 W m^{-2} is found.

However, the presented work is not to be seen as advocating this climate engineering approach. Instead, the obtained results induce a requirement for further studies. For example, a comprehensive characterization of the Arctic aerosol burden, as well as measurements and laboratory experiments for a comprehensive understanding ice nucleation, as being the particular target of this method, may help reducing uncertainties. Furthermore, the microphysical processes occurring in mixed-phase clouds are still afflicted with uncertainties and lack of understanding.

Contents

1 Introduction

Modern-day anthropogenic activities affect the atmosphere and thus global climate in many-faceted ways. The most decisive impacts here are emissions of aerosols and trace gases from various sources and with differing strength and composition. Due to economic growth in nearly all parts of the world, these pattern change and amplify since the last decades. Among the many consequences of this process is a remarkable increase in global traffic, especially in air traffic. All kind of perturbations of the atmosphere generate certain feedback mechanisms. Currently, air traffic contributes with 0.05 (0.02 – 0.15) W m^{-2} to about 2 to 8 % to the entire anthropogenic radiative forcing (Myhre et al., 2013). In several aspects, aircraft emissions differ from other sources, as they release climate-relevant substances in pristine regions and high altitudes, hence modify the concentrations of e. g. carbon dioxide (CO_2), nitric oxides, soot and water vapor. Prospectively, their importance is growing due to increasing numbers of flight movements. Additionally to the effects of such trace gases, also formation of contrails is an important aspect to consider.

1.1 Contrails and their impact on shortwave radiation and photovoltaic power production

Contrails consist of ice crystals formed in the exhaust plume of aircraft due to mixing of the hot and humid exhaust with cold environmental air. Contrails form in case, the Schmidt-Appleman-Criterion (Schmidt, 1941; Appleman, 1953; Schumann, 1996) is fulfilled, i. e. the ambient temperature is below a threshold of around $-45°$ C. With plume temperatures near $-38°$ C to

$-40°$ C, contrail particles, which are formed in the liquid phase initially, freeze homogeneously and quickly to form ice crystals. Those ice crystals grow in air with relative humidity above ice saturation and contrails persist. With such a favorable state of the atmosphere, the originally line-shaped contrails undergo various physical processes at the micro scale, spread by the influence of shear and sedimentation and change their structure and microphysical properties. At some point, contrails are no longer distinguishable from natural cirrus in observations. This type of anthropogenic cloud is then called contrail cirrus (Heymsfield et al., 2010) and can have lifetimes of several hours. In a particular case, an 18-hour old contrail-cirrus could be tracked in a satellite imagery (Minnis et al., 1998). Other examples of long-lifetime contrail observations are summarized by Schumann and Heymsfield (2017).

Contrails influence the radiative budget of the atmosphere in a way that is comparable to that of thin natural cirrus clouds (Sausen et al., 2005). Although a number of observations (e. g. Iwabuchi et al., 2012; Vázquez Navarro et al., 2015; Schumann et al., 2017) and other modeling studies have been performed, important properties, such as the optical depth or the spatial and temporal extent of occurrence, have not been sufficiently investigated and are also not quantified to a satisfactory extent (Boucher et al., 2013). The radiative forcing of aged contrails and contrail cirrus is of greater importance than the one originating from young, line-shaped contrails (Stubenrauch and Schumann, 2005; Eleftheratos et al., 2007; Burkhardt and Kärcher, 2011).

Previous model studies primarily used GCM[1]s extended by parameterizations that are able to simulate line-shaped contrails. Here, the global radiative forcing due to contrails was quantified. The values range from roughly 5 to 20 mW m^{-2} depending on the simulated year and assumed contrail properties (Marquart et al., 2003; Stuber and Forster, 2007; Kärcher et al., 2010) and are consistent with satellite-based estimates by Spangenberg et al. (2013). Taking the larger effect of contrail cirrus into account, a global mean radiative forcing

[1] Global Circulation Model

of approximately 38 mW m^{-2} (Burkhardt and Kärcher, 2011) (recently updated to 56 mW m^{-2} by Bock and Burkhardt, 2016b) to 40 to 80 mW m^{-2} (Schumann and Graf, 2013) is found. The major drawbacks of these methods are the coarse resolution of the models in both space and time.

Another class of studies simulates single contrails with high resolution LES[2] or RANS[3] models, hereby focusing on contrail formation (Paoli et al., 2013; Khou et al., 2015), young contrails with ages up to 5 minutes and their interaction with the descending wake vortices (Lewellen and Lewellen, 2001; Unterstrasser, 2014) and the transition into contrail-cirrus over time scales of hours (Unterstrasser and Gierens, 2010a; Lewellen, 2014). Usually, the contrail evolution is studied for a variety of environmental scenarios which helps to single out important process and ambient and aircraft parameters. Parameter studies allow for investigating the conditions under which contrails are persistent and the manner in which microphysical and optical properties change during transition and decay. Because this method is not applicable to a larger number of contrails and often limited to idealized environmental scenarios, it is not suited to quantify the impact of air traffic on the state of the atmosphere.

During the past decades, the development of alternative, clean energy production was enhanced to counteract global warming and reduce air pollution. Within the scope of methods, one of the most promising sources is solar energy, gained by photovoltaic (PV) power cells. To assure a sustainable supply, the demand of precise prediction of the energy yield from PV systems is desired (Lew and Richard, 2010). Several approaches exist to forecast PV power, such as statistical models, neural networks, remote sensing models and numerical weather prediction models (Inman et al., 2013). Especially PV forecast using numerical weather prediction models is challenged by special weather situations or phenomena that are poorly represented in the models (Köhler et al., 2017). E. g. Rieger et al. (2017) found a large impact of mineral

[2] Large-Eddy Simulation
[3] Reynolds-Averaged Navier Stokes

dust due to Saharan dust outbreaks on the solar radiation over Germany. This becomes important, as mineral dust is currently not considered adequately in operational weather forecast. Another phenomenon that is not represented at all in numerical weather prediction, is the influence of aviation.

In the first part of this thesis (section 3), the development of a detailed set of parameterizations for describing the life cycle of aircraft induced cloudiness in the framework of the high-resolving, regional weather prediction model COSMO-ART is described. This encompasses the online-coupled treatment of contrail microphysics, i. e. interaction with the moisture budget as well as with ice crystals in natural cirrus, and the impact of the atmospheric radiative fluxes.

Within an exemplary case study, a first attempt is started to quantify the difference of aircraft induced cloudiness compared to natural cirrus clouds. Based on this, the impact of contrails and contrail cirrus on the radiative fluxes can be examined, especially with regard to photovoltaic power production. Furthermore, the importance of one special aspect concerning contrail formation is assessed that until now was generally neglected: To what extent can ice crystal loss in young contrails determine contrail-cirrus properties? This work is based on an earlier study by Gruber (2015). Here, the technical basis was created. However, several substantial improvements were necessary to evaluate the questions raised above with full comprehensiveness. Firstly, the rather coarse approximation of the contrail formation process adapted from Schumann (2012) is replaced by a more suitable parameterization based on results of LES studies after Unterstrasser (2016). This choice also allows to assess the process of ice crystal loss in the early phase of contrail formation as being decisive for determining properties of aged contrails and contrail cirrus. Furthermore, the description of the contrail ice crystals is modified. In Gruber (2015), ice crystals in contrails were assumed to have a certain shape not changing with age. In the present study, a bimodal size distribution is used, hence more accurately considering change of ice crystal shapes with time and size.

1.2 Climate engineering by Arctic winter cirrus thinning

Air traffic is only one aspect of the anthropogenic activities influencing the atmospheric processes. A variety of global anthropogenic emissions of climate-relevant substances is contributing to a change in global climate since the beginning of industrialization. Climate change is characterized by increasing global mean temperatures and changing large-scale dynamical structures, accompanied by changes and intensification of severe weather phenomena such as droughts, flooding, or storms (Stocker et al., 2013).
Since the last years, special emphasis is being placed on this topic both in public, as well as in politics. Subsequently, in 2015, finally the Paris Agreement was signed by members of the United Nations to conduct measures for limiting the global mean temperature increase to 1.5° C referred to pre-industrial values.
However, such goals are ambitious and hard to fulfill by a reduction of emissions only. Hence, concerned climate scientists have proposed several ideas for how to compensate or at least decelerate climate warming (Crutzen, 2006; Caldeira et al., 2013). Subsequently, climate engineering started facing a growing attention, given that, in most climate projection scenarios aiming at fulfilling the Paris Agreement, so-called negative emissions need to be installed latest at the end of the 21th century (e. g. Sanderson et al., 2016). Negative emissions, i. e. carbon dioxide removal techniques (CDR), aim at an active reduction of CO_2 from the atmosphere (Caldeira et al., 2013). Such measures would directly counteract climate warming by reducing greenhouse gas radiative forcing. Several approaches are envisaged, e. g. afforestation to increase terrestrial CO_2 uptake, or measures to increase activity of marine organisms and plants consuming CO_2. However, besides the issues of reservoirs not being able to permanently store CO_2, or possible accidents leading to strong re-emissions, such measures, most likely, are unrealistically expensive (Shepherd, 2009).

Another class of climate engineering methods, in contrast to CDR not explicitly mentioned in the Paris agreement, subsumed as solar radiation management (SRM), comprises several approaches hypothetically able to alter the Earth's radiation budget, i. e. not directly tackling the source, but rather the impact of enhanced greenhouse gas concentrations (Boucher et al., 2013; Irvine et al., 2016).

One of the most prominent examples is stratospheric aerosol injection (SAI), i. e. the releasing of sulphate aerosol or suitable precursors in the stratosphere. The resulting layer of durable, reflective aerosol particles is supposed to reflect incoming solar radiation back into space and thereby cool the atmosphere (e. g. Crutzen, 2006; Kravitz et al., 2013).

Several other measures aim at altering albedo to increase reflection of incoming solar radiation by, e. g. whitening buildings, growing crops with suitable properties (Ridgwell et al., 2009; Irvine et al., 2011), or even installing space mirrors (Angel, 2006).

As clouds exert a strong influence on the Earths' energy balance (Wild et al., 2015), several methods were proposed that focus on large-scale modification of clouds to achieve surface cooling, e. g. increasing the albedo and thus the reflectivity of low-level marine stratocumulus clouds by injecting aerosol into pristine regions (Latham, 1990; Alterskjær et al., 2013; Boucher et al., 2017).

As for the CO_2 removal techniques, also most of the methods here would be extremely expensive. In addition, besides the effects aimed for, further processes might be triggered, causing harmful and undesired consequences such as changes in large-scale dynamics that in turn may alter precipitation patterns (Kravitz et al., 2013; Kleinschmitt et al., 2018). Furthermore, a sudden termination of these means could lead to unpredictable, abrupt responses (Jones et al., 2013; Parker and Irvine, 2018).

Another idea gaining growing attention was proposed by Mitchell and Finnegan (2009). Here, the outgoing longwave radiation (OLR) emitted from the Earth and the atmosphere is to be enhanced by thinning or even removing cirrus clouds in the upper troposphere.

Cirrus clouds are characterized by the absence of liquid droplets and are found at high altitudes and temperatures below 235 K (Pruppacher and Klett, 1997). Covering between 17 % (Sassen et al., 2008) and 30 % (Rossow and Schiffer, 1999) of the upper troposphere, they play a decisive role for both the water vapor budget of the upper troposphere and the Earth's energy budget (Myhre et al., 2013). Their net warming effect has been confirmed by analyzing satellite data (Hartmann et al., 1992; Chen et al., 2000; Futyan et al., 2005; Hong et al., 2016; Matus and L'Ecuyer, 2017), in-situ observations (Kienast Sjögren et al., 2016), and modeling studies (Gasparini and Lohmann, 2016). Two contributions lead to this effect: As cirrus clouds reflect incoming solar shortwave (SW) radiation, they cause a certain cooling. By absorbing longwave (LW) radiation from warm sources like the Earth's surface and the atmosphere beneath them, and re-emitting parts of it at much lower temperatures, a net warming effect results at the top of the atmosphere (TOA). As the LW warming effect more than outweighs the SW cooling, a net warming of 5 to 6 W m^{-2} results (Gasparini and Lohmann, 2016; Hong et al., 2016). Therefore, a reduction or even removal of cirrus clouds would result in a cooling effect. However, the radiative effects of cirrus clouds crucially depend on their optical depth and altitude, microphysical parameters like ice crystal number concentration, size and shape, as well as on surface temperature and reflectivity (Fusina et al., 2007; Corti and Peter, 2009; Joos et al., 2014).

Ice crystals in cirrus clouds can form by different processes. Cirrus clouds as anvils resulting from deep convection usually form from homogeneous freezing of cloud droplets. This process is governed by large vertical velocities due to convection and temperatures larger than $-60°$ C (Penner et al., 2015). Also ice crystals resulting from freezing of cloud droplets in mixed-phase clouds that get transported to higher altitudes may be the origin of large fractions of cirrus clouds (Luebke et al., 2016; Wernli et al., 2016; Voigt et al., 2017). Spontaneous freezing of supercooled liquid aerosol particles, such as sulfuric acid and sulphate solution, is called homogeneous nucleation. This process requires besides the ubiquitous solution droplets, temperatures below approximately 235 K, and supersaturation above 145 % (Kärcher and Lohmann, 2002). Homogeneous nucleation occurs in a 'burst' (Barahona and Nenes, 2009a). Depending on temperature, vertical velocity, droplet volume and vapor pressure (Koop et al., 2000), up to several thousands of small ice crystals (1 to 10 μm) are formed (Krämer et al., 2016). Due to the resulting high ice crystal number concentrations (n_{ICE}), crystals will stay small, as due to the consumption of water vapor during nucleation, their growth is limited (Ickes et al., 2015). Another process that leads to formation of ice crystals is heterogeneous ice nucleation on suitable ice nucleating particles (INP). This process is not fully understood yet, but most likely, the ice-nucleating efficiency is dependent on particle size and chemistry (Hoose and Möhler, 2012). Most effective are insoluble aerosol particles such as mineral dust, metallic particles, several biological particles, and possibly also soot (Phillips et al., 2008). Ice nucleation efficiency of particles is determined on a molecular level by local surface features acting as active sites aiding ice growth (Kiselev et al., 2017).

Heterogeneous ice nucleation requires lower supersaturation and can occur at higher temperatures than homogeneous nucleation, as the INP facilitate the phase transition necessary by lowering the energy barrier of freezing (Kärcher and Lohmann, 2003; Hoose and Möhler, 2012). In the upper troposphere, INP occur in much smaller number concentrations than the ubiquitous solution droplets, usually not exceeding 10 l^{-1} (DeMott et al., 2003). Therefore, fewer ice crystals result from heterogeneous than from homogeneous ice nucleation and ice crystals can grow to larger sizes of tens of μm (Kuebbeler et al., 2014; Kärcher et al., 2006; Barahona and Nenes, 2009a). Although INP seem to occur only sparsely, they can have a large impact on cloud formation (Gettelman et al., 2012), as both nucleation mechanisms compete for the available water vapor (Lohmann et al., 2008). With suitable low supersaturation and vertical velocity, ice crystals formed by heterogeneous nucleation can, by depleting excess water vapor, compensate for the increase of supersaturation due to updrafts and even reduce supersaturation. Thereby, homogeneous nucleation can be suppressed or even inhibited (DeMott et al., 1994). Based on the debatable (Cziczo et al., 2013) assumption that homogeneous nucleation is the dominant formation process of cirrus clouds, the idea of Mitchell and Finnegan (2009) makes use of the competition of the two nucleation mechanisms. Artificially injected INP in suitable regions could trigger heterogeneous ice nucleation and suppress homogeneous nucleation. Compared to ice crystals in natural cirrus clouds, the ice crystals would in that case occur in much lower number concentrations. As they would also be larger and could grow even further, sedimentation would remove them efficiently from the high altitudes, thus reducing their lifetime. In addition, their optical depth would be reduced (Jensen et al., 2016; Kärcher and Ström, 2003). Both would result in a smaller warming effect. Furthermore, sedimenting ice crystals effectively remove water vapor that is among the most important greenhouse gases (Lohmann and Gasparini, 2017).

Several studies using GCMs on climatic temporal scales investigated this method with partly contradicting results concerning the method's effectiveness. Hypothetically, removing all cirrus clouds from the atmosphere, could compensate more than the effect of CO_2 doubling, resulting in a net cooling effect of 5 to 6 W m^{-2} (Gasparini and Lohmann, 2016; Hong et al., 2016). However, the difference between scenarios where all cirrus clouds originate from either pure homogeneous and pure heterogeneous nucleation, respectively, is about 2 to 3 W m^{-2}, and thus compensates for at least 50 to 80 % of a CO_2 doubling (Lohmann et al., 2008; Storelvmo et al., 2013). Assuming a background, where 50 % of all mineral dust acts as INP, Storelvmo and Herger (2014) introduced artificial INP becoming active at 105 % supersaturation.

Limiting the seeding to regions with temperatures lower than 235 K results in a significant negative effective radiative forcing with a maximum value of -2 W m^{-2} when seeding with 18 particles per liter. Global average surface air temperatures were reduced by 1.4 K. Limiting the seeding to winter polar regions, and thereby impact only 15 % of the Earth, turns out to be sufficient, as here, homogeneous nucleation is dominant.

In addition, due to reduced or lacking solar radiation, the net radiation effect is maximized (Storelvmo et al., 2014).
However, Penner et al. (2015) found only about -0.33 ± 0.21 W m^{-2} with seeding concentrations of 20 l^{-1}. They attribute the large differences compared to former studies to different assumptions made. In their simulations, secondary organic aerosol acts as INP and deposition of water vapor onto preexisting ice crystals during nucleation is considered. Also the sub-grid scale updraft velocity is not limited. Likewise, Gasparini and Lohmann (2016) do not find a significant cooling effect due to seeding. In that study, in contrast to others, ice crystal sizes decrease with seeding. This effect more than outweighs the reduction in n_{ICE}. They conclude that ice crystals formed by seeding consume enough excess water vapor to cause decreasing sizes of the remaining crystals. In addition, seeding leads to artificial cirrus cloud formation in supersaturated but cloud-free regions. This effect gains importance, when extraordinarily effective INP are used.
Another type of studies used a prescribed, increased sedimentation velocity of ice crystals as proxy for seeding (Muri et al., 2014; Jackson et al., 2016; Crook et al., 2015). This approach can reproduce surface climate responses to seeding, but is unable to capture the changes in microphysical processes that occur when considering seeding aerosol in a cirrus formation parameterization (Gasparini et al., 2017). Applying the latter approach, a large fraction of the seeding effect is outweighed by a reduction in liquid clouds in response to increased convective activity. In addition, redistribution of ice to lower levels leads to a warming effect via mixed-phase clouds.
A crucial assumption within all of the approaches mentioned is the fraction of homogeneous and heterogeneous nucleation in the unseeded atmosphere. Several studies point at homogeneous nucleation being the dominant process in cirrus formation (Spichtinger et al., 2003, 2004; Mitchell et al., 2011). However, in-situ measurements indicate that, at least over certain regions, n_{ICE} is lower than expected from homogeneous nucleation, indicating that heterogeneous ice nucleation is the main freezing mechanism (Cziczo et al.,

2013; Jensen et al., 2013; Krämer et al., 2016).

The studies mentioned above exclusively used GCMs to investigate the effects of seeding. As vertical wind velocity (w) determines supersaturation, a realistic distribution of w is crucial for obtaining realistic fractions of heterogeneous and homogeneous nucleation (Pruppacher and Klett, 1997). However, due to the coarse resolution in most studies using GCMs, w remains a large source of uncertainty. In order to still obtain information about the unresolved variability, a common method is characterizing the sub-grid distribution of vertical velocity, by its standard deviation (σ_w) (Kalesse and Kollias, 2013; Jensen et al., 2013; Barahona and Nenes, 2011). Many modeling studies find homogeneous nucleation dominating over heterogeneous nucleation at low temperatures and high values of w (Haag and Kärcher, 2004; Jensen and Pfister, 2004; Spichtinger and Gierens, 2009; Jensen et al., 2010; Barahona and Nenes, 2011). One thus can conclude, overestimation in the parameterized sub-grid scale w is a possible source of discrepancy between models and observations (Spichtinger and Krämer, 2013; Sheyko et al., 2015; Shi and Liu, 2016; Zhou et al., 2016). Applying a high-resolution GCM, Barahona et al. (2017) find σ_w being below 0.1 m s^{-1} to a larger extent that former studies with coarse resolution. This leads to a lower frequency of homogeneous nucleation and decreasing n_{ICE} with decreasing temperature and fits better to field measurements. Although strong vertical velocities may occur at high altitudes with low temperatures, small sub-grid scale variance of w limits the frequency of homogeneous nucleation, leading on average to lower n_{ICE} (Barahona et al., 2017).

Cirrus seeding in former studies was performed in a straightforward manner, by either prescribing an increased sedimentation as proxy (e. g. Muri et al., 2014), or assuming a 'perfect' seeding scenario with regard to the ice-nucleation properties of the INP, the concentrations and the injection into altitudes most susceptible for seeding (e. g. Storelvmo et al., 2013; Gasparini et al., 2017).

Injecting aerosol into a pristine environment is a massive perturbation of the atmospheric processes. Hence, extensive consequences on large-scale dynamics and the hydrological circle are to be expected. Therefore, besides searching for configurations that achieve the highest cooling, it is of at least equal importance to emphasize possible undesirable side effects of seeding cirrus clouds. If too much seeding aerosol is injected, cirrus clouds get optically thicker than natural ones, leading to a warming effect (Storelvmo and Herger, 2014). The same occurs, when the seeding INP are present in regions dominated by heterogeneous ice nucleation and they cannot reduce supersaturation enough to prevent nucleation on natural INP (Kärcher et al., 2006). However, in certain ways, cirrus seeding seems to be less invasive than several SRM techniques. For example, SAI would result in strong cooling of the Earth's surface. However, simultaneously, also the global hydrological cycle would be affected in non-desirable ways. An overall reduction in precipitation could cause increasing atmospheric stability and changes in large-scale dynamic patterns (Kravitz et al., 2013). Such effects are found to be smaller with cirrus seeding, as here, the impact on solar radiation would be weaker (Kristjánsson et al., 2015). Nevertheless, various studies find divergent side effects of seeding, ranging from an enhancement of the hydrological circle (Kristjánsson et al., 2015) to changes in large-scale dynamics and a moderate reduction in precipitation (Storelvmo et al., 2014). Also teleconnections in terms of large-scale circulation changes in remote unseeded regions are possible (Muri et al., 2014). Further side effects, such as the impact of deposed seeding aerosol on the surface albedo, especially possible impacts on sea-ice, or the influence on the maritime ecosystems, are not quantified at all. In addition, from a technical point of view, implementing such measures would be, most likely, difficult and expensive, as large amounts of seeding material would need to be produced and deported into remote regions and high altitudes. However, before even thinking about technical issues, it is inevitable to gain confidence about the method's efficacy and side effects, both intended and unintended ones, to prevent impetuous deployment.

The second part of this thesis is dedicated to a close investigation of issues and uncertainties regarding this climate engineering method. Based on the findings of Storelvmo et al. (2014), the method is assumed to be applied solely in the hibernal Arctic. Here, due to reduced or entirely lacking solar radiation, the net effect of thinning cirrus clouds is thought to be largest. The answer on whether thinning of cirrus clouds works as intended, is closely connected to the question of the prevailing processes leading to cirrus formation. This climate engineering method only yields the desired results, in case cirrus clouds during the Arctic winter mainly form via homogeneous nucleation, mainly due to absence of natural aerosol particles. Hence, one key question in the framework of this topic is: How are the natural aerosol species distributed during the Arctic winter; and how does this determine the fraction of homogeneous and heterogeneous ice nucleation both in space and time?

A decisive aspect of thinning cirrus clouds is the seeding material that needs to be available in the right altitudes and in suitable concentrations. Up till now, the aspect of injection or transport of the seeding material was generally neglected; either suitable particles were assumed to be in place where necessary, or the entire seeding process was even approximated by increased sedimentation velocities of ice crystals in cirrus clouds. An important part of this study therefore aims at gaining deeper insight into the life cycle of the seeding aerosol in the atmosphere, ranging from emissions by suitable means, transport and sedimentation into the target area, and ice nucleation in cirrus- and mixed-phase clouds.

Closely connected to such a realistic distribution of the seeding aerosol is the investigation of side effects, in particular with respect to the impact on mixed-phase clouds that might be partly desired or rather undesired.

Former modeling studies were performed assessing climate engineering by thinning of cirrus clouds, however, exclusively making use of climate models, partly with coarse simplifications. Therefore, it is of interest to perform a detailed analysis of the processes taking place using a comprehensive, high-resolved model setup in order to review the previously drawn conclusions. In addition, when undertaking such kind of research, a very important task is revealing possible sources of uncertainty with regard to current knowledge.

To assess these questions, the online coupled modeling system ICON-ART is applied. In section 4, a process study with simulations on a cloud-resolving grid spacing over a limited area of the hibernal Arctic is presented. Here, further insight is gained into the nucleation process, cirrus properties, and changes of the latter due to seeding. Furthermore, the effect of seeding aerosol and seeded ice crystals on mixed-phase clouds is investigated. Finally, the impact of cloud properties modified in such a way on the radiative fluxes can be examined. In addition, considering a specific method of injection, a rough estimate on how much seeding aerosol would be required to attain the desired effects, can be given.

In section 5, these findings are extrapolated to the entire Arctic during winter in order to warrant their robustness. Here, it is important to learn about the reliability of seeding cirrus clouds in terms of possible failure during certain episodes. Considering both a longer time span and a larger area, it is possible to give a more reliable estimate of the impact on the radiative fluxes.

2 The model systems

In this section, the model systems used for the case studies in sections 3, 4, and 5 are briefly described. In addition, the parts of both model systems decisive and modified for the studies are presented in more detail. For a comprehensive overview, the reader is referred to several important articles and model descriptions.

2.1 COSMO

The non-hydrostatical atmospheric model system COSMO[1] is applied for a case study of high-resolved simulations of contrails over Central Europe in section 3. As being a regional-scale and high-resolving model system, COSMO, in detail described in Baldauf et al. (2011), has been applied as operational weather forecasting system with several European weather services, in particular at the German Weather Service (DWD).

The model solves the primitive thermo-hydrodynamical equations describing a compressible flow in the moist atmosphere (Doms et al., 2015b). Being a limited area model, the discretization is done on generalized rotated geographical coordinates with with a terrain-following, time-independent height coordinate. The horizontal grid structure is an Arakawa C-grid; for the vertical, a Lorenz grid is chosen. The equation system is discretized with the method of finite differences, the time integration is realized with the third order Runge-Kutta scheme.

[1] COnsortium for Small scale MOdelling

Compared to global model systems, COSMO can easily be operated on small, cloud-resolving and convection-permitting scales of only a few kilometers or less. Like this, lots of physical processes can be resolved explicitly. Especially with regard to processes depending crucially on vertical velocity and the small-scale fluctuations of the latter, horizontally high resolved model systems are in great advantage compared to GCMs that are usually operated on coarser grids. Several studies pointed at the correlation of resolved vertical velocity and horizontal grid spacing (e. g. Rauscher et al., 2016; Barahona et al., 2017). As both maximum values as well as the distribution of vertical velocity agrees better with measurements for higher resolutions, the high-resolved COSMO simulations are a valuable tool to gain insight i. e. into the processes leading to cloud formation.
Instead of the operationally one moment bulk microphysical scheme, the model system can be coupled with a comprehensive two moment scheme after Seifert and Beheng (2006).
The atmospheric radiative fluxes in the COSMO model are calculated using the GRAALS (General Radiative Algorithm Adapted to Linear-type Solutions) radiation scheme (Ritter and Geleyn, 1992).

2.2 ICON

The non-hydrostatical, online-coupled model system ICON[2] is applied for assessing the impact of artificially inject aerosol into the Arctic hibernal UTLS region to modify cirrus clouds in sections 4 and 5. ICON consists of a non-hydrostatical dynamical core and packages of physical parameterizations suitable for numerical weather prediction, climate modeling as well as large-eddy simulations (Zängl et al., 2015; Heinze et al., 2017).
The following is based on Zängl et al. (2015); for further details, the reader is referred to this article. The discretization is performed on an unstructured icosahedral-triangular Arakawa C grid. The base of the grid generation in

[2] ICOsahedral Non-hydrostatic

ICON is a icosahedron projected on a sphere. In the so-called root division step, the great circle arcs that form the edges of the preliminary grid are divided into n parts. Connecting the new edge points result in n^2 new triangles within the original one. The new triangles are divided via k into bisections. The resulting grid then is termed as *Rn Bk*. The effective grid spacing (Δx) can be calculated from the average area of the grid cells (A_c) or based on the number of root divisions and bisections based on the Earth's radius (r_E):

$$\Delta x = \sqrt{A_c} = \sqrt{\frac{\pi}{5} \frac{r_E}{n 2^k}} \tag{2.1}$$

ICON offers the possibility to locally refine the grid spacing and the corresponding integration time step. Like this, within one simulation, besides a global simulation, also high-resolved areas of special interest can be examined using the same set of physical parameterizations. Furthermore, ICON also can be operated in a so called limited area mode. Unlike global modeling systems and similar to COSMO, only a certain area is selected as simulation domain, and lateral boundary conditions for meteorological variables are updated hourly. This allows the conduction of a large number of computational relatively inexpensive simulations.
Like COSMO, also ICON can be optionally coupled with the two moment scheme after Seifert and Beheng (2006).
The radiative fluxes in ICON are calculated using the Rapid Radiative Transfer Model (RRTM) (Mlawer et al., 1997).

2.3 Aerosols dynamics

The ART[3] module is an extension to both COSMO and ICON comprising a detailed treatment of aerosol dynamics and gas-phase chemistry and the associated feedback processes on both microphysics and radiative fluxes (Vogel

[3] Aerosols and Reactive Trace gases

et al., 2009; Rieger et al., 2015). Although both from a technical, as well as from a physical point of view, ART is not identical for COSMO and ICON, the brief description within this section is valid for both versions.
The species considered within ART are described on the same grid and integrated using the same numerical methods and time steps applied in COSMO or ICON for the hydrometeor species. Furthermore, ART is online coupled to the atmospheric models, i. e. the ART routines are used within the actual COSMO or ICON simulation, thus inducing a prognostic feedback on cloud formation and the radiative fluxes, hence avoiding a diagnostic description of the interactions.
Throughout the case studies presented in the subsequent sections, the aspect of gas-phase chemistry is not considered, hence formation of secondary organic aerosol and nucleation of aerosol from gas-phase precursor species are neglected. Instead, as the main focus is put on the formation of ice crystals, the ubiquitous presence of sulphate aerosol droplets (Köhler and Seifert, 2015) is approximated by assuming values for number concentration and size being constant in both time and space.
In sections 4 and 5, the impact of an artificial seeding aerosol on clouds, originally formed due to the presence of realistic aerosol distributions, is examined. Hence only mineral dust and sea salt aerosol, as being the most prevalent aerosol species for activation of cloud droplets, or rather nucleation of ice crystals, are considered within these studies. The particle size distribution of both mineral dust and sea salt aerosol are approximated by three log-normal distributions, called *modes* hereafter, associated with prognostic number concentrations and mass mixing ratios. Keeping the standard deviations σ_{i} constant during the simulation, the modes' mean diameters ($d_{0,\mathrm{i}}$ and $d_{3,\mathrm{i}}$) are diagnostic variables being able to change due to aerosol dynamical processes like sedimentation, dry deposition and washout. An overview of the initial values for modal parameters of both aerosol species can be found in Tab. 2.1.

Table 2.1: Initial parameters of mineral dust and sea salt aerosol. $d_{0,\mathrm{i}}$ and $d_{3,\mathrm{i}}$ are the median diameters of the specific number and mass emission for mode *i*; σ_i is the standard deviation for mode *i*.

mineral dust	mode *A*	mode *B*	mode *C*
$d_{0,\mathrm{i}}$ in μm	0.6445	3.454	8.672
$d_{3,\mathrm{i}}$ in μm	1.5	6.7	14.2
σ_i	1.7	1.6	1.5
sea salt	mode *A*	mode *B*	mode *C*
$d_{0,\mathrm{i}}$ in μm	0.2	2.0	12.0
$d_{3,\mathrm{i}}$ in μm	0.69	8.5	27.9
σ_i	1.9	2.0	1.7

A detailed description of parameterizations for mineral dust emission can be found in Rieger et al. (2017). The emission fluxes of sea salt aerosol are calculated online depending on wind speed and temperature, hence considering various sources like film, jet and spume droplets (Lundgren et al., 2013).

2.4 Cloud microphysics

As already mentioned before, both COSMO and ICON operationally make use of a well proven single moment bulk microphysical scheme to calculate the processes related to cloud formation and precipitation. However, both model systems may be coupled with a more comprehensive, but computationally more expensive two moment scheme after Seifert and Beheng (2006).

Just like with the ART extension (section 2.3), the scheme is not exactly the same for both model systems regarding technical and physical details. Nevertheless, the basic equations and methods are similar. For the brief overview given here, it therefore seems sufficient to describe the principal and decisive methods, both versions have in common. All formulae can be found and explained in greater detail Seifert and Beheng (2006). Values of constants used in this section are listed in Tab. A.1 and Tab. A.2 in appendix A.The scheme considers six hydrometeor classes (cloud droplets, ice, rain, snow, graupel, hail) that are described by prognostic budget equations for both mass and number concentrations. In principle, the scheme prognostically describes size distributions of hydrometeors. Size distributions can be characterized by the zeroth and the first moment, of which the zeroth is the specific number, whereas the first is proportional to the mass mixing ratio. The particle size distribution of each class is characterized by a generalized Γ-distribution (Eq. 2.2).

$$f(m) = A\, m^{\nu} \exp\left[-\lambda m^{\mu}\right] \tag{2.2}$$

Here, m is the particle mass in units of kg. The parameters ν and μ (see Tab. A.1 in appendix A) are assumed being constant, respectively.
Expressions involving Γ-functions exist for the moments of order k (Eq. 2.3).

$$M^k(m) = \int_0^{\infty} m^k f(m)\, \mathrm{d}m = n\overline{m}^k \left[\frac{\Gamma\left(\frac{\nu+1}{\mu}\right)}{\Gamma\left(\frac{\nu+2}{\mu}\right)}\right]^k \frac{\Gamma\left(\frac{\nu+1+k}{\mu}\right)}{\Gamma\left(\frac{\nu+1}{\mu}\right)} \tag{2.3}$$

The moment of order $k = 0$ is the number density n of the distribution $f(m)$, whereas the first moment corresponds to the mass density q (Eq. 2.4).

$$n = \int_0^{\infty} f(m)\, \mathrm{d}m\ ; \quad q = \int_0^{\infty} m f(m)\, \mathrm{d}m \tag{2.4}$$

Using these relations, the parameters A and λ can be defined as in Eq. 2.5 and Eq. 2.6. The parameter A is proportional to the number concentration n, whereas λ is related to the particles' mean mass $\overline{m} = q/n = M^1/M^0$.

$$A = n\mu \frac{\lambda^{\frac{\nu+1}{\mu}}}{\Gamma\left(\frac{\nu+1}{\mu}\right)} \tag{2.5}$$

$$\lambda = \left[\frac{\Gamma\left(\frac{\nu+1}{\mu}\right)}{\Gamma\left(\frac{\nu+2}{\mu}\right)} \overline{m} \right]^{\mu} \tag{2.6}$$

The shape of particles in each class is approximated using a power-law formulation (Eq. 2.7). The mean mass m (in kg) of a single particle is related to its size D (in meters) via a mass-size relation with coefficients a_{geo} and b_{geo} for each hydrometeor class given in Tab. A.1 in appendix A.

$$D = a_{\mathrm{geo}} m^{\mathrm{b_{geo}}} \tag{2.7}$$

Using relations relying on power law functions like Eq. 2.7 is beneficial. Then integrals over the mass distribution can often be expressed in terms of moments which avoids employing more expensive numerical quadrature techniques.
The budget equation solved for the kth Moment of each hydrometeor class j follows Eq. 2.8.

$$\frac{\partial M_j^k}{\partial t} + \nabla \cdot \left(\vec{v} M_j^k\right) - \nabla \cdot \left(K_h \nabla M_j^k\right) + \frac{\partial}{\partial z}\left(\overline{v}_{\mathrm{sed,j,k}} M_j^k\right) = S_j^k \tag{2.8}$$

Here, $\vec{v}$ is the three dimensional wind velocity, K_{h} is the turbulent diffusion coefficient, $\overline{v}_{\mathrm{sed}}$ is the sedimentation velocity, and $S^{\mathrm{k}}{}_{\mathrm{j}}$ represents sinks and sources. Henceforth, the index j is refers to *liquid clouds* (c), *rain* (r), *ice* (i), *graupel* (g), or *hail* (h). The zeroth moment describing the number density is abbreviated with n_{j}, the first moment, i. e. the mass density, is abbreviated with q_{j}.
For the liquid phase, the last term of Eq. 2.8 consists of accretion, autoconversion, self-collection and break up, whereas frozen hydrometeors undergo diffusional growth, freezing, aggregation, self-collection, riming and melting.

As the following case studies mainly focus on processes occurring within the ice-phase, the parameterizations of the most important processes are described here. A more thorough description of all processes involved can be found in Seifert and Beheng (2006).
Amongst the most important sources is the process of nucleation, i. e. the formation of new particles from the gas phase. This process depends crucially on the properties of the aerosol species involved. Therefore, a more detailed explanation is deferred to section 2.5.

Growth of ice crystals by water vapor deposition

The growth equation of a single ice crystal is given by (Pruppacher and Klett, 1997).

$$\frac{dm}{dt} = \frac{4\pi C(m) F_{\text{ven}}(m) s_{\text{i}}}{\frac{R_{\text{V}}T}{p_{\text{sat,i}}(T)D_{\text{V}}} + \frac{L_{\text{iv}}}{K_{\text{T}}T}\left(\frac{L_{\text{iv}}}{R_{\text{V}}T} - 1\right)} = 4G_{\text{iv}}(T,p)\, s_{\text{i}} D(m) F_{\text{ven}}(m) \tag{2.9}$$

Here, T is the temperature, p is the pressure, and $C = D/\pi$ denotes the capacity of hexagonal crystals (Harrington et al., 1995) with D following Eq. 2.7. s_{i} is the supersaturation with respect to ice, L_{iv} represents the latent heat of sublimation, $p_{\text{sat,i}}$ denotes the saturation vapor pressure over ice, R_{v} is the specific gas constant for water vapor, K_{T} is the conductivity of heat, and D_{v} is the molecular diffusion coefficient of water vapor. The term F_{ven} accounts for ventilation effects and G_{iv} considers the diffusion of water vapor and the effect of latent heating (Eq. 2.10).

$$G_{\text{iv}}(T,p) = \left[\frac{R_{\text{v}}T}{p_{\text{sat,i}}D_{\text{v}}} + \frac{L_{\text{iv}}}{K_{\text{T}}T}\left(\frac{L_{\text{iv}}}{R_{\text{v}}T} - 1\right)\right]^{-1} \tag{2.10}$$

Integration of Eq. 2.9 over the ice crystal mass spectrum yields the temporal derivative of the ice mass density q_{i}.

$$\begin{aligned}\frac{\partial q_{\mathrm{i}}}{\partial t} &= 4G_{\mathrm{iv}}(T,p)\,s_{\mathrm{i}}\int_{0}^{\infty} D(m)\,F_{\mathrm{ven}}(m)\,f(m)\,\mathrm{d}m \\ &= 4G_{\mathrm{iv}}(T,p)\,s_{\mathrm{i}}D(\overline{m})\,\overline{F}_{\mathrm{ven}}\end{aligned} \tag{2.11}$$

Here, $\overline{F}_{\mathrm{ven}}$ is the averaged ventilation coefficient.

Interaction of hydrometeor classes

For the *ice class* containing all frozen hydrometeors too small for being classified as snow, graupel or hail, the mass-size relation given by Eq. 2.7 uses the values $a_{\mathrm{geo}} = 0.835$ and $b_{\mathrm{geo}} = 0.39$ (see Tab. A.1 in appendix A). In order to avoid unreasonably small or large mean masses $\overline{m} = q_{\mathrm{i}}/n_{\mathrm{i}}$, a lower limit $m_{\mathrm{min}} = 10^{-12}$ kg and upper limit $m_{\mathrm{max}} = 10^{-6}$ kg are introduced. If $\overline{m}$ lies outside the interval $[m_{\mathrm{min}}, m_{\mathrm{max}}]$ in a grid box, the ice crystal number concentration is increased to $q_{\mathrm{i}}/m_{\mathrm{min}}$ or reduced to $q_{\mathrm{i}}/m_{\mathrm{max}}$, respectively. The limits correspond to sizes $L_{\mathrm{min}} = 17.5\ \mu$m and $L_{\mathrm{max}} = 3800\ \mu$m. The same procedure, however with different threshold values (see Tab. A.1 in appendix A), is applied for all classes of hydrometeors considered.

In the model, both self-collection and collection between the individual classes of frozen hydrometeors are considered. In case that two different hydrometeor classes are present, where class A has a larger mean mass than class B, the ice crystals of class B are collected by those of class A. In addition to the number loss in class B, mass is transferred from class B to class A.

Furthermore, ice and snow particles can be converted to graupel by riming of cloud droplets. This happens, when the collected mass during riming fills up a sphere related to the maximum diameter of the original ice crystal, hence a critical rime mass for the conversion from ice to graupel is defined in Eq 2.12.

$$m_{\text{crit}} = \alpha_0 \rho_{\text{w}} \left(\frac{\pi}{6} \overline{D}^3 - \frac{\overline{m}}{\rho_{\text{i}}} \right) \tag{2.12}$$

Here, $\overline{D}$ is the mean diameter and $\overline{m}$ the mean mass of the ice crystal class, $\rho_{\text{w}} = 1000$ kg m^{-3} and $\rho_{\text{i}} = 900$ kg m^{-3} are the densities of water and ice, respectively, and the so-called space-filling coefficient $\alpha_0 = 0.68$. The characteristic time τ_{rim} is defined in Eq 2.13.

$$\tau_{\text{rim}} = m_{\text{crit}} \left[\frac{\partial \overline{m}}{\partial t} \right]^{-1} = \alpha_0 \rho_{\text{w}} n \left(\frac{\pi}{6} \overline{D}^3 - \frac{\overline{m}}{\rho_{\text{i}}} \right) \left[\frac{\partial q_{\text{i}}}{\partial t} \right]^{-1} \tag{2.13}$$

With $\partial q_{\text{i}} / \partial t$ being the mass flux of cloud droplets onto the ice crystals. Finally, the conversion rate $\partial q_{\text{g}} / \partial t$ to the graupel class can be calculated as the ratio of mass density of the riming ice crystals and the characteristic time (Eq. 2.14).

$$\frac{\partial q_{\text{g}}}{\partial t} = \frac{q_{\text{i}}}{\tau_{\text{rim}}} = \frac{\partial q_{\text{i}}}{\partial t} \left[\alpha_0 \frac{\rho_{\text{w}}}{\rho_{\text{i}}} \left(\frac{\frac{\pi}{6} \rho_{\text{i}} \overline{D}^3}{\overline{m}} - 1 \right) \right]^{-1} \tag{2.14}$$

The conversion rate of graupel number concentration $\partial n_{\text{g}} / \partial t$ (Eq. 2.15) is obtained by dividing Eq. 2.14 by the mean ice crystal size.

$$\frac{\partial n_{\text{g}}}{\partial t} = \frac{1}{\overline{m}} \frac{\partial q_{\text{g}}}{\partial t} \tag{2.15}$$

The same procedure is applied for the snow-to-graupel conversion assuming $\alpha_0 = 0.01$.

Sedimentation

Similar to the mass-size relation, the terminal sedimentation velocity v_{sed} is approximated by a power law (Eq. 2.16) with coefficients a_{vel} and b_{vel} given in Tab. A.2.

$$v_{\text{sed}}(m) = a_{\text{vel}} m^{b_{\text{vel}}} \tag{2.16}$$

Integrating Eq. 2.16 over the particle size distribution assumed to follow a Γ-distribution yields mean fall velocities $\overline{v}_{\text{sed,k}}$ for the kth moment.

$$\overline{v}_{\text{sed,k}} = a_{\text{vel}} \frac{\Gamma\left(\frac{k+\nu+b_{\text{vel}}+1}{\mu}\right)}{\Gamma\left(\frac{k+\nu+1}{\mu}\right)} \left[\frac{\Gamma\left(\frac{\nu+1}{\mu}\right)}{\Gamma\left(\frac{\nu+2}{\mu}\right)}\right]^{b_{\text{vel}}} \overline{m}^{\text{b}_{\text{vel}}} \tag{2.17}$$

In Eq. 2.16 and Eq. 2.17, m and $\overline{m}$ are in kg, v_{sed} and $\overline{v}_{\text{sed,k}}$ are in in m s^{-1}.

Numerical treatment of the microphysical parameterizations

By construction, the two moment microphysical scheme is designed for applications with high horizontal resolution accompanied by small integration time steps, usually not exceeding 25 s (Seifert and Beheng, 2006). In order to obtain realistic process rates also when used for global ICON-ART simulations with a rather coarse horizontal grid spacing of about 40 km and an integration time step of 150 s, several modifications are necessary.
The main issue occurs with sedimentation of large hydrometeors, especially graupel and hail that have large sedimentation velocities and thus are removed from the atmosphere too fast for interacting in physical reasonably manner with other hydrometeors. This results e. g. in very high number concentrations of ice crystals and cloud droplets that otherwise would have been collected or could have rimed onto the graupel and hail particles. Additionally, with

small distances between model layers close to the surface, the Courant number may exceed one when large hydrometeors sediment too fast. Circumvent this by simply limit the sedimentation flux in an explicit scheme results in too high mass and number concentrations of the hydrometeors in the lowest layers. Also, applying an implicit solver is not reasonable due to the non-linear dependence of the sedimentation velocity from mass- and number mixing ratio (Gaßmann, 2002).

To correct for this behavior, two possible solutions are presented here.

The first is introducing a simple dynamical sub-cycling of the sedimentation process, i. e. the sedimentation flux of the larger hydrometeors is consecutively calculated several times with a shorter integration time step. Both the number of iterations as well as the integration time step are chosen according to the maximum vertical distance between two adjacent grid layers, preventing the hydrometeors to fall deeper than one grid layer per iteration step. The number of iterations (n_{sed}) is determined using Eq. 2.18.

$$n_{\mathrm{sed}} = v_{\mathrm{sed,max}} \frac{\Delta t}{\Delta z} \tag{2.18}$$

Here, $v_{\mathrm{sed,max}}$ is the maximum allowed sedimentation velocity of the current hydrometeor species, Δt is the global integration time step, and Δz is the distance between the current adjacent grid layers. The integration time step used for sub-cycling then is $\Delta t / n_{\mathrm{sed}}$.

Like this, at least numerical stability can be achieved, as the Courant number is not being violated in the bottom layers. However, still no meaningful interaction of the rapidly sedimenting large hydrometeor species with smaller ones can be described. Therefore, a quasi-implicit solver making use of a predictor-corrector method is applied to include the term containing sources and sinks (${S^{\mathrm{k}}}_{\mathrm{j}}$ in Eq. 2.8) into the sedimentation process. A more detailed derivation of this this formulation that also is used in the COSMO model, can be found in Gaßmann (2003); Doms et al. (2015a).

For any specific moment M, i. e. number or mass density of a single hydrometeor class, rewriting Eq. 2.8 by conflating all terms not solved within the microphysics, i. e. advection and turbulent diffusion, yields Eq. 2.19.

$$\frac{\partial M}{\partial t} = R - \frac{\partial}{\partial z}(\overline{v}_{\text{sed}} M) + S \tag{2.19}$$

Here, R incorporates advection and turbulent diffusion, the second term represents the sedimentation, S contains all microphysical source and sink processes. Like most atmospheric model systems, ICON uses a process time-splitting method (Zängl et al., 2015). Following this, the 'dynamics' (term R in Eq. 2.19) are integrated over a time step first. The resulting preliminary values for the prognostic variable M serve as input for the 'physics', in this case consisting of sedimentation and S. During this second step, the prognostic variables are updated finally. Hence, during the 'physical' step, Eq. 2.19 reads as in Eq. 2.20.

$$\frac{\partial M}{\partial t} = -\frac{\partial}{\partial z}(\overline{v}_{\text{sed}} M) + S \tag{2.20}$$

The implicit, mass-conserving Crank-Nicolson discretization of Eq. 2.20 can be expressed according to Eq. 2.21.

$$\phi^1{}_k = \phi^0{}_k + \frac{\Delta t}{2\Delta z_k}\left(P^1{}_{k-\frac{1}{2}} - P^1{}_{k+\frac{1}{2}} + P^0{}_{k-\frac{1}{2}} - P^0{}_{k+\frac{1}{2}}\right) + S_k \Delta t \tag{2.21}$$

Here, the partial density of the hydrometeor class $\phi = M$, the sedimentation flux is $P = \overline{v}_{\text{sed}} M$. Values superscribed with *0* are values obtained form the 'dynamical' step, values superscribed with *1* denote final values. The index k counts the model layers from top to bottom.

A formulation using a Leapfrog integration scheme with one-sided differences results from Eq. 2.21 when applying an upstream approximating for flux at half-levels $k \pm 1/2$ and estimating the sedimentation velocities from ϕ at the adjacent full levels $k \pm 1$ (Eq. 2.22).

$$\phi^1{}_k = \phi^0{}_k + \frac{\Delta t}{2\Delta z_k} \Delta_1 (\phi, v) + S_k \Delta t \tag{2.22}$$

with

$$\Delta_1 (\phi, v) = \phi^1{}_{k-1} v^1{}_{k-1} - \phi^1{}_k v^1{}_k + \phi^0{}_{k-1} v^0{}_{k-1} - \phi^0{}_k v^0{}_k \tag{2.23}$$

The unknown value $v^1{}_\mathrm{k}$ in (Eq. 2.23) is replaced by an average of the already known ϕ at the current level k and the one above $(k-1)$.

$$v^1{}_k = v \left(\frac{1}{2} \phi^1{}_{k-1} + \frac{1}{2} \phi^0{}_k \right) \tag{2.24}$$

Using Eq. 2.24, Eq. 2.22 reads as:

$$\phi^1{}_k = y_k \left(\phi^0{}_k + \frac{\Delta t}{2\Delta z_k} \Delta_2 (\phi, v) + S_k \Delta t \right) \tag{2.25}$$

Here, $\Delta_2 (\phi, v)$ reads as:

$$\Delta_2 (\phi, v) = \phi^1{}_{k-1} v^1{}_{k-1} + \phi^0{}_{k-1} v^0{}_{k-1} - \phi^0{}_k v^0{}_k \tag{2.26}$$

In Eq. 2.25, an implicit weight as defined in Eq. 2.27 is introduced.

$$y_k = \left[1 + \frac{\Delta t {v^1}_k}{2\Delta z_k}\right]^{-1} \tag{2.27}$$

Negative values for ϕ in Eq. 2.25can be avoided by introducing a flux limiter as defined in Eq. 2.28.

$$\phi^1{}_k = y_k\left(\phi^0{}_k + \mathrm{MAX}\left[\frac{\Delta t}{2\Delta z_k}\Delta_2(\phi, v), -\phi^0{}_k\right] + S_k\Delta t\right) \tag{2.28}$$

For each grid column, Eq. 2.28 is solved explicitly for each model layer. A predictor-corrector approach is applied to account for microphysical interactions during sedimentation in such way that Eq. 2.28 is solved neglecting the very last term, hence producing predictor values $\phi^*{}_\mathrm{k}$.

$$\phi^*{}_k = y_k\left(\phi^0{}_k + \mathrm{MAX}\left[\frac{\Delta t}{2\Delta z_k}\Delta_2(\phi, v), -\phi^0{}_k\right]\right) \tag{2.29}$$

Results obtained from Eq. 2.29 serve as input for the calculation of the actual microphysical processes.

2.5 Aerosol-cloud interaction

The indirect aerosol effect summarizes various processes involving aerosol particles modifying cloud properties and thus have an indirect effect on the Earth's radiation budget (e. g. Myhre et al., 2013).

The most important processes are the activation of aerosol particles to act as cloud condensation nuclei (CCN), homogeneous nucleation, i. e. freezing of supercooled liquid aerosol droplets, and heterogeneous nucleation summarizing various nucleation processes with solid aerosol particles involved.

To obtain realistic rates for the formation of ice crystals, a careful description of the competition between homogeneous and heterogeneous nucleation of ice crystals is required. This section briefly reviews the basic principles realized in the parameterization after Barahona and Nenes (2009b) used in ICON-ART.
The very basis of nucleation theory is the temporal evolution of supersaturation with respect to ice (s_i) within an ascending air parcel in Eq. 2.30 after Pruppacher and Klett (1997).

$$\frac{\mathrm{d}s_\mathrm{i}}{\mathrm{d}t} = \alpha w (1 + s_\mathrm{i}) - \beta \frac{\mathrm{d}q_\mathrm{i}}{\mathrm{d}t} \tag{2.30}$$

$$\alpha = \frac{L_\mathrm{iv} M_\mathrm{w} g}{c_\mathrm{p} R T^2} - \frac{g M_\mathrm{a}}{RT} \tag{2.31}$$

$$\beta = \frac{p M_\mathrm{a}}{p_\mathrm{sat,i} M_\mathrm{w}} - \frac{{L_\mathrm{iv}}^2 M_\mathrm{w}}{c_\mathrm{p} R T^2} \tag{2.32}$$

Here, q_i is the mass mixing ratio of ice, and the molar mass of water and air, M_w and M_a, respectively. The second term in Eq. 2.30, $\beta \frac{\mathrm{d}q_\mathrm{i}}{\mathrm{d}t}$, reduces supersaturation, as it represents a sink term of water vapor due to condensation. Instead, $\alpha w (1 + s_\mathrm{i})$ is a source of supersaturation, as the adiabatic cooling due to the rising motion increases s_i.
As mentioned above, homogeneous and heterogeneous nucleation are the two major formation processes determine the formation of ice crystals in the atmosphere.
Homogeneous freezing requires very high supersaturation and cannot take place with temperatures above $-38°$ C (Pruppacher and Klett, 1997). This process depends only on temperature, supersaturation with respect to ice and the particle size distribution of the aerosol population, hence it is independent from the actual chemical composition of the aerosol (Koop et al., 2000). In case, supersaturation exceeds a certain threshold s_crit of about 40 %, nucleation sets in. Instantaneously, due to the usually high concentrations of liquid aerosol droplets, numerous small ice crystals form. Due to the high nucleation rate,

supersaturation is reduced rapidly; homogeneous nucleation virtually occurs in a 'burst' (Barahona and Nenes, 2009a). Additionally, condensation aids in reducing supersaturation. Therefore, the maximum supersaturation possible s_{max} hardly exceeds s_{crit}. Due to the short time scale of this process, it needs to be parameterized in ICON-ART. The fraction of nucleated ice particles is approximated by scaling the constant nucleation rate at the maximum supersaturation $J(s_{max}) = 10^{16}$ s^{-1}m^{-3} after Koop et al. (2000).
With temperatures around $-10°$ C, supercooled water can exist. In case of supersaturation with respect to water, condensation freezing, i. e. freezing while condensation of water on the aerosol particle takes place. Freezing of the already established liquid coat around an aerosol particle is called immersion freezing. Contact freezing describes the collision of a solid particle with a supercooled water droplet resulting in instantaneous freezing of the conglomerate. Finally, deposition freezing gives name to the formation of ice crystals on the surface of insoluble aerosol particles. This process requires temperatures lower than $-20°$ C but may occur below supersaturation with respect to water (Hoose and Möhler, 2012).

Parameterization of ice nucleation

The complex process of heterogeneous nucleation is parameterized based on observations after Phillips et al. (2013). Here, besides temperature, supersaturation and the aerosols' size distribution, also both their surface area and chemical composition are considered. The formulation distinguishes between different aerosol species such as organic aerosol, soot and mineral dust.
The competition between homogeneous and heterogeneous nucleation concerning the amount of excess environmental water vapor is realized via the definition of a limiting IN concentration (n_{LIM}) after Barahona and Nenes (2009b). In case, n_{LIM} is reached, s_{max} is lower than s_{crit}, hence, homogeneous nucleation is inhibited entirely. Below this threshold, pure homogeneous

nucleation can occur, and provided that IN are present at all, always a combined homogeneous and heterogeneous freezing will occur. The resulting number of ice crystals formed (n^*) hence can be expressed as the sum of homogeneous (n_{hom}) and heterogeneous (n_{het}) nucleated ice crystals in Eq. 2.33.

$$\mathrm{n}^* = \begin{cases} n_{\text{hom}} + n_{\text{het}}\left(s_{\text{crit}}\right) & n_{\text{het}}\left(s_{\text{crit}}\right) < n_{\text{LIM}} \\ n_{\text{het}}\left(s_{\max}\right) & n_{\text{het}}\left(s_{\text{crit}}\right) \geq n_{\text{LIM}} \end{cases} \tag{2.33}$$

A last crucial step within the parameterization of nucleation of ice crystals is the calculation of a nucleation rate $J_{\text{n,NUC}}$ that can be used as source term $S_0{}^{\text{i}}$ in the budget equation (Eq. 2.8). A common approach is correcting the parameterized number of nucleated ice crystals n^* with the number of already existing ones (n_{ICE}) (e. g. Seifert and Beheng, 2006) for avoiding unrealistically high number concentrations caused by repetitive calculations of the nucleation parameterization (Eq. 2.34).

$$J_{\text{n,NUC}} = \text{MAX}\left[\frac{n^* - n_{\text{ICE}}}{\Delta t}, 0\right] \tag{2.34}$$

The corresponding rate for the nucleated ice mass $J_{\text{q,NUC}}$ then simply follows from Eq. 2.35.

$$J_{\text{q,NUC}} = m_{\text{NUC}}\, J_{\text{n,NUC}} \tag{2.35}$$

Here, $m_{\text{NUC}} = 10^{-12}$ kg is the minimum mean mass of ice crystals.

Parameterization of sub-grid scale vertical velocity

As seen in Eq. 2.30, supersaturation crucially depends from vertical velocity w. However, even with a relatively high spatial resolution of 5 km, parts of the dynamical structures decisive for cloud formation remain unresolved (Rauscher et al., 2016; Barahona et al., 2017), the sub-grid scale variability of the vertical velocity is approximated according to Morales and Nenes (2010), hence applying a Gaussian distribution (Eq. 2.36) centered around the value of the grid-scale vertical velocity w. Firstly, this method was described for use in COSMO-ART by Bangert et al. (2012).

$$P_w(w') = \frac{1}{\sqrt{2\pi\sigma_{\text{TKE}}}} \exp\left[\frac{(w'-w)^2}{2\sigma_{\text{TKE}}^2}\right] \tag{2.36}$$

To account for the sub-grid scale variability of the vertical velocity, the standard deviation σ_{TKE} is based on the turbulent kinetic energy (TKE) (Eq. 2.37) after Lohmann et al. (1999), as turbulence, besides gravity waves, in general dominate the small-scale variability of atmospheric vertical movements (Lohmann et al., 1999; Morales and Nenes, 2010).

$$\sigma_{\text{TKE}} = f_{\text{w}}\sqrt{TKE} \tag{2.37}$$

For use in coarse-resolution global models, Lohmann et al. (1999) proposed $f_{\text{w}} = 0.7$, whereas Rieger et al. (2015) found $f_{\text{w}} = 0.3$ yielding better results in higher resolution ICON-ART simulations of a mid-latitude spring episode compared to measurements.

Using this approach, the mathematically correct formulation of the nucleation parameterization then reads as:

$$\overline{n}^* = \frac{\int_0^\infty n^*\left(w', [...]\right) P_{\text{w}}(w')\, \mathrm{d}w'}{\int_0^\infty P_{\text{w}}(w')\, \mathrm{d}w'} \tag{2.38}$$

Numerically, Eq. 2.38 is approximated by weighting the results of the parameterization evaluated at several sampling points of P_{w}.

On the parameterization of cloud droplet formation

For the formation of cloud droplets, a parameterization based on classical activation theory is used (Fountoukis and Nenes, 2005). This parameterization calculates the activation of log-normally distributed aerosol species at the maximum supersaturation possible. The latter is determined on the diabatical cooling rate, dependent on vertical movement and thermodynamical state. The decelerated growth of giant cloud condensation nuclei (Barahona et al., 2010) is considered as well as the activation of insoluble particles via Frenkel-Halsey-Hill adsorption theory (Kumar et al., 2009). When using this configuration with a two-moment scheme, it is important to distinguish whether activation takes place either within a preexisting cloud or at its base or rather forming an entirely new cloud (Bangert et al., 2012).

2.6 Calculation of the cloud optical properties

In this section, the impact of clouds on the radiative fluxes in the atmosphere are discussed. In a first, highly simplifying step, the impact on the radiative fluxes crossing a single cloud layer containing spherical particles is derived according to (Petty, 2004).

For reasons of consistency, a certain doubling of variable abbreviations occurs due to the non-uniform conventions within the literature.

According to Beer's law, the transmittance t of cloud layer is defined as in Eq. 2.39.

$$t = \exp\left[-\frac{\tau}{\cos(\theta)}\right] \tag{2.39}$$

Here, τ is the optical thickness of the cloud layer, θ is the solar zenith angle. The optical thickness is defined as extinction coefficient β_{ext} integrated over the entire vertical extension of the cloud layer reaching from z_1 to z_2 (Eq. 2.40).

$$\tau = \int_{z_1}^{z_2} \beta_{\mathrm{ext}}(z)\,\mathrm{d}z \tag{2.40}$$

The extinction coefficient is defined in Eq. 2.41. For simplicity, a monodisperse distribution of spherical particles with radius r_{p} and mass density ρ_{p} is assumed.

$$\beta_{\mathrm{ext}} = n_{\mathrm{p}} \sigma_{\mathrm{ext}} \pi {r_{\mathrm{p}}}^2 = k_{\mathrm{ext}} q_{\mathrm{p}} = k_{\mathrm{ext}} n_{\mathrm{p}} \frac{3}{4} \pi {r_{\mathrm{p}}}^3 \rho_{\mathrm{p}} \tag{2.41}$$

Here, σ_{ext} is the extinction efficiency, hence the ratio of extinction cross section and geometric cross section, n_{p} is the particle number concentration within the cloud, q_{p} is the mass concentration, and k_{ext} is the mass specific extinction coefficient. The latter is defined as:

$$k_{\mathrm{ext}} = \frac{\beta_{\mathrm{ext}}}{q_{\mathrm{p}}} = \frac{3\sigma_{\mathrm{ext}}}{4\rho_{\mathrm{p}} r_{\mathrm{p}}} \tag{2.42}$$

Within the visible spectrum, using $\sigma_{\mathrm{ext}} = 2$ for spherical particles is a valid approximation (Petty, 2004). Inserting Eq. 2.41 and Eq. 2.42 into Eq. 2.40 yields Eq. 2.43. For reasons of convenience, in the following, liquid clouds are considered, hence $p = c$. Nonetheless, the formulations can universally be applied to all kind of particles.

$$\tau = \frac{3}{2\rho_{\mathrm{i}} r_{\mathrm{i}}} LWP \tag{2.43}$$

With the liquid water path LWP defined as the vertically integrated mass concentration (Eq. 2.44).

$$LWP = \int_{z_1}^{z_2} q_\mathrm{c}\,\mathrm{d}z \tag{2.44}$$

Using Eq. 2.44, Eq. 2.43 reads as:

$$\tau = \left[\frac{9LWP^2\pi\,(z_2 - z_1)}{2\rho_\mathrm{c}{}^2} n_\mathrm{c}\right]^{\frac{1}{3}} \tag{2.45}$$

Assuming both LWP and the vertical extension of the cloud layer being constant, τ is proportional to $n_\mathrm{c}{}^{1/3}$, which is a good approximation for liquid clouds.

However, in reality, hydrometeors are not monodispersely distributed, hence the presented derivation needs to be redone integrating over an assumed particle size distribution $f(r)$ depending on particle radius r.

Doing so, Eq. 2.41 now reads as:

$$\beta_\mathrm{ext} = \int_0^\infty f(r)\,\sigma_\mathrm{ext}\pi r^2\,\mathrm{d}r \tag{2.46}$$

$$q_\mathrm{i} = \int_0^\infty f(r)\,\rho_\mathrm{c}\frac{4}{3}\pi r^3\,\mathrm{d}r \tag{2.47}$$

For simplicity, the particles are still assumed to be of spherical shape. Dividing Eq. 2.46 by Eq. 2.47 yields τ, still assuming $\sigma_\mathrm{ext} = 2$.

$$\tau = \frac{\beta_{\text{ext}}}{q_{\text{c}}} LWP = \frac{\int\limits_0^\infty f(r)\, \sigma_{\text{ext}} \pi r^2 \, \mathrm{d}r}{\int\limits_0^\infty f(r)\, \rho_{\text{c}} \frac{4}{3} \pi r^3 \, \mathrm{d}r} LWP \approx \frac{3}{2\rho_{\text{c}}} \frac{LWP}{r_{\text{e}}} \tag{2.48}$$

In Eq. 2.48, r_{e} is the effective radius of the size distribution. In general, it is defined as in Eq. 2.49.

$$r_{\text{e}} = \frac{\int\limits_0^\infty f(r)\, r^3 \, \mathrm{d}r}{\int\limits_0^\infty f(r)\, r^2 \, \mathrm{d}r} \tag{2.49}$$

As radiative transfer algorithms usually require the extinction coefficient β_{ext}, rearranging Eq. 2.48 results in Eq. 2.50.

$$\beta_{\text{ext}} \approx \frac{3 q_{\text{c}}}{2 \rho_{\text{c}} r_{\text{e}}} \tag{2.50}$$

Eq. 2.50 is valid for the extinction of of solar radiation for a single cloud layer. However, in the real atmosphere, incoming solar radiation gets scattered several times, also the approximation of a single cloud layer usually is not valid. In order to account for such effects, a three dimensional solution of the radiative transfer equation requires information about the scattering angles and the contribution of absorption to calculation both direct and diffuse, i. e. scattered, radiative fluxes.

The latter can be approximated using a two-stream approach, hence distinguishing between upward and downward diffuse fluxes in a single column assuming isotropy throughout the atmosphere. Like this, the modification of the radiative fluxes due to clouds and other obstacles such as gases or aerosol

particles is characterized by the extinction coefficient, the single scattering albedo, and the asymmetry factor (Petty, 2004).
The single scattering albedo ω is the the fraction of scattering within the extinction coefficient (Eq. 2.51).

$$1-\omega=\frac{\beta_{\text{abs}}}{\beta_{\text{ext}}} \tag{2.51}$$

Here, β_{abs} is the absorption coefficient, defined as in Eq. 2.52.

$$\beta_{\text{abs}}=\int_{0}^{\infty}\sigma_{\text{abs}}f(r)\,\mathrm{d}r \tag{2.52}$$

With the absorption cross section σ_{abs}.
The asymmetry factor g is the first moment of the scattering phase function, characterizing the relative strength of the forward scattering. For practical reasons, g is defined as the average of the cosine of the scattering angle obtained form a high number of scattering events (Petty, 2004).
To approximate the entire solar and thermal spectrum, the optical properties need to be known for several wavelength bands, depending on the size distribution of the hydrometeors or the aerosol population.
The atmospheric radiative fluxes in the COSMO-ART model are calculated using the GRAALS (General Radiative Algorithm Adapted to Linear-type Solutions) radiation scheme (Ritter and Geleyn, 1992). The radiative fluxes are calculated using a two-stream approximation of the radiative transfer equations for eight spectral bands (Tab. B.2 in appendix B).
In ICON-ART, the radiative fluxes are calculated using the Rapid Radiative Transfer Model (RRTM) (Mlawer et al., 1997) based on the correlated k approach. The radiative transfer equations are solved, as with GRAALS, using a two-stream approximation for 30 wavelength bands(Tab. B.1).

Neglecting the differences concerning both the technical implementation as well as with physical details between the two approaches, however, the representation of gases, aerosols, and clouds are fairly similar.
In the standard configuration, in both schemes, optical properties of gases as well as of aerosols are calculated based on climatological profiles. However, when either COSMO or ICON are coupled with the ART extension, the optical properties of various aerosol species are parameterized depending on the prognostic spatial and temporal distributions, as well as on the simulated size distributions. The interactions of mineral dust and sea salt with the radiative fluxes realized in ICON-ART are described in detail in Rieger et al. (2017) and Gasch et al. (2017); a comprehensive description of the implementation in COSMO-ART can be found in Bangert et al. (2012).

2.6.1 A consistent description of the effective radius

In both GRAALS and RRTM, the impact of hydrometeors on the radiative fluxes is approximated deriving cloud optical properties based on the effective radii r_e (in ICON following Stevens et al. (2013)). This formulation only considers the mass mixing ratio of cloud ice and cloud droplets, respectively designed to be used together with a single moment microphysical scheme.
Cloud optical properties required by both schemes are the extinction coefficient, the single scattering albedo, and the asymmetry factor.
A first attempt to implement an improved description of cloud optical properties has been conducted by U. Blahak, H. Muskatel, and P. Khain in order to obtain a more realistic representation of liquid cloud optical properties within the COSMO model using the standard one-moment bulk microphysical scheme.
As throughout the studies presented here, a two moment microphysical scheme is applied, detailed and more comprehensive description of cloud microphysical properties is available. Especially, information on the particle size distributions are used here, to derive more accurate cloud optical

properties.
Hence, within the framework of these studies, the approach mentioned above was extended to be applied together with a two moment microphysical scheme and especially to be used also for frozen hydrometeors.
For a better representation of particle shape, size and number concentration, the formulation of r_e after Fu (1996), and the mean axis ratio A_r after Fu (2007), both based on mean particle size or mass, respectively, for cloud droplets and ice crystals is applied.
The formulation for the effective radius r_e for ice crystals after Fu et al. (1998) with diameter D and length L and particle size distribution $f(L)$ is given in Eq. 2.53.

$$r_e = \frac{1}{2} \frac{\int_0^\infty D^2 L f(L)\, \mathrm{d}L}{\int_0^\infty f(L) \left(DL + \frac{\sqrt{3}}{4} D^2 \right) \mathrm{d}L} \tag{2.53}$$

For using Eq. 2.53 that is formulated in dependence of the mean particle length, together with the size distributions used in the microphysical parameterizations (section 2.4) $f(m)$, hence depending on mean particle mass m, certain manipulations are necessary. A more detailed derivation can be found in appendix B. The final expression of $r_e(m)$ is given in Eq. 2.54 with coefficients c_1 to c_4 in Eqs. 2.55.

$$r_e(m) = \frac{1}{2} \left[c_1 m^{c_2} + c_3 m^{c_4} \right]^{-1} \tag{2.54}$$

$$
\begin{aligned}
c_1 &= \sqrt{\frac{3\sqrt{3}\rho a_{\text{geo}}}{8} \frac{\Gamma\left(\frac{\nu+\frac{3}{2}-\frac{b_{\text{geo}}}{2}}{\mu}\right)}{\Gamma\left(\frac{\nu+2}{\mu}\right)} \left[\frac{\Gamma\left(\frac{\nu+1}{\mu}\right)}{\Gamma\left(\frac{\nu+2}{\mu}\right)}\right]^{\frac{1}{2}(1-b_{\text{geo}})}} \\
c_2 &= \frac{1}{2}\left(1-b_{\text{geo}}\right) \\
c_3 &= \frac{\sqrt{3}}{4a_{\text{geo}}} \frac{\Gamma\left(\frac{\nu+2-b_{\text{geo}}}{\mu}\right)}{\Gamma\left(\frac{\nu+2}{\mu}\right)} \left[\frac{\Gamma\left(\frac{\nu+1}{\mu}\right)}{\Gamma\left(\frac{\nu+2}{\mu}\right)}\right]^{b_{\text{geo}}} \\
c_4 &= b_{\text{geo}}
\end{aligned}
\tag{2.55}
$$

The mean axis ratio

In contrast to the parameterization of other cloud optical properties, formulations of the asymmetry factor g depending on r_e are sensitive to the assumed habit, i. e. columns or plates, and shape of ice crystals (Fu, 1996). Subsequently, Fu (2007) found the ice crystals' mean axis ratio A_r to be the quantity for obtaining more robust parameterizations of g.

The mean axis ratio of a single ice crystal is defined in Eq. 2.56 after (Fu, 2007).

$$
A_r = \frac{\int_0^\infty \frac{D}{L} A_p f(L)\, dL}{\int_0^\infty A_p f(L)\, dL} \tag{2.56}
$$

In Eq. 2.56, A_p is the statistically average projected area of a particle that is, as random orientation of the particles is assumed, $1/4$ of its total surface area S.

$$A_\mathrm{p} = \frac{1}{4}S = \frac{3}{4}\left(\frac{\sqrt{3}}{4}D^2L + DL\right) \tag{2.57}$$

$$A_\mathrm{r}(m) = \frac{c_1 m^{c_2} + c_3 m^{c_4}}{c_5 m^{c_6} + c_7 m^{c_8}} \tag{2.58}$$

$$\begin{aligned}
c_1 &= \frac{2}{3\rho} a_\mathrm{geo}{}^{-\frac{3}{2}} \Gamma\left(\frac{\nu + \frac{5}{2} - \frac{3b_\mathrm{geo}}{2}}{\mu}\right) \left[\frac{\Gamma\left(\frac{\nu+1}{\mu}\right)}{\Gamma\left(\frac{\nu+2}{\mu}\right)}\right]^{-\frac{1}{2}(5-3b_\mathrm{geo})} \\
c_2 &= -\frac{1}{2}\left(5 - 3b_\mathrm{geo}\right) \\
c_3 &= \frac{1}{\sqrt{a_\mathrm{geo}}} \Gamma\left(\frac{\nu + \frac{3}{2} - \frac{b_\mathrm{geo}}{2}}{\mu}\right) \left[\frac{\Gamma\left(\frac{\nu+1}{\mu}\right)}{\Gamma\left(\frac{\nu+2}{\mu}\right)}\right]^{-\frac{1}{2}(3-b_\mathrm{geo})} \\
c_4 &= -\frac{1}{2}\left(3 - b_\mathrm{geo}\right) \\
c_5 &= \frac{1}{\sqrt{2\sqrt{3}\rho}} \Gamma\left(\frac{\nu+2}{\mu}\right) \left[\frac{\Gamma\left(\frac{\nu+1}{\mu}\right)}{\Gamma\left(\frac{\nu+2}{\mu}\right)}\right]^{-2} \\
c_6 &= -2 \\
c_7 &= \sqrt{a_\mathrm{geo}} \Gamma\left(\frac{\nu + \frac{3}{2} + \frac{b_\mathrm{geo}}{2}}{\mu}\right) \left[\frac{\Gamma\left(\frac{\nu+1}{\mu}\right)}{\Gamma\left(\frac{\nu+2}{\mu}\right)}\right]^{-\frac{1}{2}(3+b_\mathrm{geo})} \\
c_8 &= -\frac{1}{2}\left(3 + b_\mathrm{geo}\right)
\end{aligned} \tag{2.59}$$

In a similar way as described for r_e, Eq. 2.56 needs to be rearranged to depend from m instead of L. Details on the derivation can be found in appendix B.

The final expression is given in Eq. 2.58, with coefficients c_1 to c_8 listed in Eq. 2.59.

Effective radius of cloud droplets

For cloud droplets, the effective radius is parameterized as the ratio between second and third moment (Eq. 2.60).

$$r_e = \frac{1}{2}\frac{M^3}{M^2} = \frac{1}{2}\frac{\int_0^\infty D^3 f(D)\,\mathrm{d}D}{\int_0^\infty D^2 f(D)\,\mathrm{d}D} \tag{2.60}$$

For consistency, also Eq. 2.60 is rearranged to depend on m:

$$r_e = c_1 m^{c_2} \tag{2.61}$$

with c_1 and c_2:

$$\begin{aligned} c_1 &= \frac{1}{2} a_{geo} \frac{\Gamma\left(\frac{\nu+1+3b_{geo}}{\mu}\right)}{\Gamma\left(\frac{\nu+1+2b_{geo}}{\mu}\right)} \left[\frac{\Gamma\left(\frac{\nu+1}{\mu}\right)}{\Gamma\left(\frac{\nu+2}{\mu}\right)}\right]^{b_{geo}} \\ c_2 &= b_{geo} \end{aligned} \tag{2.62}$$

2.6.2 Revised cloud optical properties

Consequently, also the optical properties of cloud droplets and ice crystals are revised. In this section, a brief description of the procedure is given, based on the work of U. Blahak, H. Muskatel, and P. Khain.

For ice crystals, β_{ext}, ω, and g are parameterized as function of r_e and A_r, respectively. The fits are based on a composite of the single scattering

parameters extinction efficiency (σ_{ext}), scattering efficiency (σ_{ext}), and asymmetry factor (g) obtained from Mie theory, anomalous diffraction theory, geometric optics method, and finite-difference time domain technique covering wavelengths ranging from 0.21 μm to 100 mm (Fu, 1996).In the following, only an overview of the fitting procedure for is given. A more detailed explanation is given in appendix B.
First, the basic definition of the single scattering properties are integrated over a particle size distribution to obtain spectral optical properties. For this, size distributions of ice crystals are assumed to follow a generalized Γ-distribution depending on the mean crystal size L.

$$f(L) = AL^{\nu} \exp\left[-\lambda L^{\mu}\right] \tag{2.63}$$

The formulations for the spectral optical properties are defined as following.

$$\beta_{\text{ext}} = \int_{0}^{\infty} \sigma_{\text{ext}}(L) f(L) \, \mathrm{d}D \tag{2.64}$$

$$\beta_{\text{sca}} = \int_{0}^{\infty} \sigma_{\text{sca}}(L) f(L) \, \mathrm{d}D \tag{2.65}$$

$$\beta'_{\text{ext}} = \frac{\beta_{\text{ext}}}{IWC} \tag{2.66}$$

$$\omega = \frac{\beta_{\text{sca}}}{\beta_{\text{ext}}} \tag{2.67}$$

$$\omega = \frac{\int_0^\infty g' \sigma_{\text{sca}}(L) f(L) \, \mathrm{d}D}{\int_0^\infty \sigma_{\text{sca}}(L) f(L) \, \mathrm{d}D} \; ; \; g' = \frac{1}{2} \int_{-1}^{1} \cos\theta P(\cos\theta) \, \mathrm{d}\cos\theta \tag{2.68}$$

with scattering angle θ and the scattering phase function P.
First, a particle size distribution (PSD) ensemble is defined by systematic variations of the PSD parameters within a wide range. Then, for each PSD of this ensemble, the scattering parameters are computed and spectrally averaged over the wavelength bands (λ) of the corresponding radiation scheme. Averaging β_{ext} is done after Eq. 2.69.

$$\overline{\beta'_{\text{ext}}} = -\frac{1}{\rho_i \Delta z} \ln \left[\frac{\int_{\lambda_{\min}}^{\lambda^{\min}} S(\lambda) \exp\left[-\beta'_{\exp} \rho_i \Delta z\right] \, \mathrm{d}\lambda}{\int_{\lambda_{\min}}^{\lambda^{\min}} S(\lambda) \, \mathrm{d}\lambda} \right] \tag{2.69}$$

Here, S is the Planck-function, ρ_{i} is the density of ice, and Δz is the model layer thickness.
In the same way, $\overline{\omega}$ (Eq. 2.70) and $\overline{g}$ (Eq. 2.71) are averaged .

$$\overline{\omega} = \frac{\int_{\lambda_{\min}}^{\lambda_{\max}} S(\lambda) \beta_{\text{ext}} \omega \, \mathrm{d}\lambda}{\int_{\lambda_{\min}}^{\lambda_{\max}} S(\lambda) \beta_{\text{ext}} \, \mathrm{d}\lambda} \tag{2.70}$$

$$\overline{g} = \frac{\int_{\lambda_{\min}}^{\lambda_{\max}} S(\lambda) \beta_{\text{sca}} g \, \mathrm{d}\lambda}{\int_{\lambda_{\min}}^{\lambda_{\max}} S(\lambda) \beta_{\text{sca}} \, \mathrm{d}\lambda} \tag{2.71}$$

Finally, β_{ext} and ω are parameterized as function of r_{e}, and g as function of A_{r} by applying rational, non-linear functions (Eq. 2.72).

$$f(x) = \frac{\sum_{i=0}^{N} a_i x^i}{\sum_{i=0}^{M} b_i x^i} \tag{2.72}$$

For RRTM, the resulting parameterizations are Eq. 2.73, Eq. 2.74, and Eq. 2.75, with cofficients listed in Tab. C.7, Tab. C.8, and Tab. C.9.

$$\beta_{\mathrm{ext}} = \frac{c_0 + c_1 2r_{\mathrm{e}} + c_2 (2r_{\mathrm{e}})^2}{c_3 + c_4 2r_{\mathrm{e}} + c_5 (2r_{\mathrm{e}})^2 + c_6 (2r_{\mathrm{e}})^3} \tag{2.73}$$

$$\omega = \frac{c_0 + c_1 2r_{\mathrm{e}} + c_2 (2r_{\mathrm{e}})^2 + c_3 (2r_{\mathrm{e}})^3}{c_4 + c_5 2r_{\mathrm{e}} + c_6 (2r_{\mathrm{e}})^2 + c_7 (2r_{\mathrm{e}})^3} \tag{2.74}$$

$$g = \frac{c_0 + c_1 A_{\mathrm{r}} + c_2 {A_{\mathrm{r}}}^2 + c_3 {A_{\mathrm{r}}}^3}{c_4 + c_5 A_{\mathrm{r}} + c_6 {A_{\mathrm{r}}}^2 + c_7 {A_{\mathrm{r}}}^3 + c_8 {A_{\mathrm{r}}}^4} \tag{2.75}$$

For GRAALS, the resulting parameterizations are of the same as for RRTM, except for g (Eq. 2.76). Fit coefficients are listed in Tab. C.1, Tab. C.2, and Tab. C.3.

$$g = \frac{c_0 + c_1 A_{\mathrm{r}} + c_2 {A_{\mathrm{r}}}^2 + c_3 {A_{\mathrm{r}}}^3}{c_4 + c_5 A_{\mathrm{r}} + c_6 {A_{\mathrm{r}}}^2 + c_7 {A_{\mathrm{r}}}^3} \tag{2.76}$$

Likewise, for cloud droplets, data derived from Mie theory (Hu and Stamnes, 1993) is employed. Here, in contrast to the procedure described for ice crystals, fitting functions of a power-law-type (Eq. 2.77) have proven yield better results. Furthermore, to overcome the lower and upper size limits of 5 μm and 60 μm, respectively, of the original data, the fits are performed for three size regions, assuring that the fit function for β_{ext} asymptotically approaches zero for $r_e \rightarrow \infty$, whereas for ω, r_e is simply limited to 100 μm.

$$f(r_e) = a_1 r_e{}^{a_2} + a_3 \tag{2.77}$$

The resulting coefficients for the wavelength bands used in GRAALS are listed in appendix C in Tab. C.4, Tab. C.5, and Tab. C.6; for RRTM in Tab. B.2 and Tab. C.10, Tab. C.12, and Tab. C.11.

Although the presented derivations are based on the assumption of hexagonal ice crystals, or rather cloud droplets, the formulations of r_e and A_r are generally uniform and not limited to a certain type of hydrometeor nor their shape. Therefore, they presented description of the cloud optical properties can be extended to also consider rain, snow and graupel in the radiative transfer algorithms.

3 Contrails and their impact on shortwave radiation and photovoltaic power production - a regional model study

In this study, the impact of contrails on the shortwave radiation and subsequently on the production of photovoltaic power are investigated. In large parts, this chapter is based on an article titled as *Contrails and their impact on shortwave radiation and photovoltaic power production - a regional model study* published in *Atmospheric Chemistry and Physics* (Gruber et al., 2018). The author of this thesis is also the lead author of the article. However, valuable contributions to the article of Simon Unterstrasser, Jan Bechtold, Heike Vogel, Martin Jung, Henry Pak, and Bernhard Vogel are gracefully acknowledged.

The spatial scale applied in this study lies between those typically used for large-eddy simulations and global climate models using the online coupled regional-scale model system COSMO-ART. In this context, online coupled means that meteorology, chemistry and contrail-related processes are simulated in one model at the same grid and one main time step for integration is used. For this study, the model is extended by introducing a new hydrometeor class for contrail ice crystals. The contrail initialization is based on a parametrization of early contrail properties by Unterstrasser (2016). The prognostic equations for contrail ice are similar to those of the natural cirrus ice class. Despite their similar treatment in terms of model equations, the evolution of the two cloud types can be quite different, as their formation mechanisms are different. Especially with respect to the spatial scale used in the regional scale COSMO-

ART model, the presented study is complementary to the aforementioned approaches and is, the first study of its kind.
Ice crystals in young contrails are considerably smaller than the ones that form in natural cirrus clouds and occur with substantially higher ice crystal number concentrations (Febvre et al., 2009; Schröder et al., 2000; Voigt et al., 2010). Therefore, the original microphysical scheme was extended by a new hydrometeor class that allows for a separate treatment of the small contrail ice crystals separate from natural ice. This approach allows for the investigation of contrail microphysical properties and their changes during the various stages of development represented in a regional atmospheric model. In contrast to other studies using a two-moment microphysical scheme (Bock and Burkhardt, 2016a), the presented model configuration uses no fractional cloud coverage and no prognostic equations for contrail geometric properties like volume or area which describe the contrail spreading on the subgrid scale. Compared to GCM parameterizations, this omission seems acceptable in a regional model, as the spatial resolution is much higher (horizontal grid size of 2.8 km versus 50 km in a GCM).
Regarding the contrail microphysics and the interaction with the meteorological situation, the entire procedure is online coupled, thus allowing feedback processes between contrails and natural clouds in contrast to other models on a comparable grid scale (Schumann, 2012). There, a mixed Lagrangian-Eulerian approach is used instead of the usual Eulerian treatment. This approach allows covering the scale ranging from thousands of single contrails to multi-year global climate simulations (Schumann, 2012; Schumann et al., 2015; Caiazzo et al., 2017).
One of the key goals of this study is thus to quantify the influence of contrails and contrail cirrus on natural high-level cloudiness. Moreover, the radiative properties of contrails and their local influence on the shortwave radiative fluxes at the surface are examined.
The model uses a diagnostic radiation scheme (Ritter and Geleyn, 1992). Because the description of shortwave optical properties for ice clouds is

optimized for various naturally occurring crystal habits (Fu et al., 1998; Key et al., 2002), a separate treatment of contrail ice crystals is introduced here as well.
Another feature of this study is the new and recently compiled data set of flight trajectories. Rather than statistical calculations for globally averaged fuel consumption, the basic data consist of real commercial aircraft waypoint data (flightradar24.com, 2015). Another approach using commercial flight data, to study contrails on a regional scale is described in Duda et al. (2004). Here, a combination of commercial flight data and coincident meteorological satellite remote sensing data was used to perform a case study of a widespread contrail cluster. In the future, further case studies should be performed for which in-situ observations of natural ice clouds and especially contrails are available.
The presented model configuration serves to study microphysical evolution of contrails and contrail cirrus, their influence on natural high-level cloudiness, and their impact on the radiative fluxes on a regional scale and short time periods. This gains importance, e.g., in predicting the energy yield from photovoltaic systems.
In section 3.1, the modification of the microphysical scheme and the radiation scheme are described. Section 3.1.2 presents a description of the parameterization providing the source terms for contrail ice. In section 3.2, the results of a case study and comparison with satellite observations are presented.

3.1 The contrail parameterizations

In this study, the COSMO-ART model is coupled with the two-moment microphysical scheme following Seifert and Beheng (2006). Until now, the scheme contained one cloud ice class that describes ice crystals in high-level ice clouds. Because ice crystals in freshly formed contrails are considerably smaller than those in natural cirrus, the basic microphysical processes in young contrails are treated in a separate, newly introduced *contrail ice class*. From now on, the original ice cloud class is called *cirrus ice class*. The separate treatment allows simulating local bi-modal size spectra.

3.1.1 The new contrail ice class

In this section, the relevant contrail-specific modifications implemented into the microphysical parameterizations in order to represent contrails are presented. However, in large parts, the treatment of contrail ice is analogous to that of natural cirrus (see section 2.4). Thus, unless otherwise indicated, the parameterized processes for the new contrail ice class are the same as those used for the cirrus ice class.

Similar approaches with a separate contrail ice class using climate models with coarser grid size are described by Burkhardt and Kärcher (2009) and Bock and Burkhardt (2016a).

In contrast to the natural ice class, that assumes constant shape factors for all mass range considered in Eq. 2.7, for the contrail ice class, a piecewise definition of a_{geo} and b_{geo} is employed following (Spichtinger and Gierens, 2009; Heymsfield and Iaquinta, 2000). For masses above $m_{\mathrm{split}} = 2.15 \times 10^{-13}$ kg (corresponds to $L_{\mathrm{split}} = 7.42\ \mu$m), $a_{\mathrm{geo,con}} = 4.25$ and $b_{\mathrm{geo,con}} = 0.45$ are used. For masses below m_{split}, $a_{\mathrm{geo,con}} = 0.124$ and $b_{\mathrm{geo,con}} = 0.33$ is prescribed which defines quasi-spherical hexagonal columns with aspect ratio 1 (see derivation in Spichtinger and Gierens, 2009). The latter constants are valid down to the prescribed lower limit $m_{\mathrm{min}} = 10^{-15}$ kg. For grid boxes with lower mean masses, the same bounding technique is used as in the cirrus ice class. The upper limit is set to a relatively small value of $m_{\mathrm{max}} = 2 \times 10^{-11}$ kg and the treatment of grid boxes with too large mean masses is different compared to the cirrus ice class. Instead of bounding n, the total ice crystal mass and number from such a grid box are transferred from the contrail ice class to the cirrus ice class. The prescribed mass limits of contrail ice class correspond to the size limits $L_{\mathrm{min}} = 1.24\ \mu$m and $L_{\mathrm{max}} = 58\ \mu$m.

The contrail ice class is involved into the same physical processes as the natural ice class. However, nucleation is switched off in the contrail ice class. Instead, the generation of contrail ice depends on air traffic and the atmospheric state and is explained in section 3.1.2.

As mentioned before, most ice crystals in contrails are smaller than in natural cirrus and different values for a_{geo} and b_{geo} (given in Tab. A.3 in appendix A) assure reasonable aspect ratios, also for small ice crystals down to sizes of $L_{\min} = 1.24\ \mu$m. The upper size limit is set to a relatively small value of $L_{\max} = 58\ \mu$m. If the mean ice crystal size in a grid box exceeds $L_{\max}$, then the total ice crystal mass and number from such a grid box are transferred from the contrail ice class to the cirrus ice class. This is reasonable, as contrails show distinct bi-modal size spectra with many small ice crystals with sizes around 10 μm and fewer large ice crystals in the fall streaks (Unterstrasser et al., 2017a; Lewellen et al., 2014). The contrail ice class contains predominantly small ice crystals and the cirrus ice class allows for larger ice crystals that may also stem from aged contrails or contrail fall streaks. One drawback of this approach is that the anthropogenic contribution in the cirrus ice class cannot be directly determined. Instead, we analyze the differences in the cirrus ice class between simulations with and without air traffic. This indirect quantification of the aged contrail contribution could be circumvented by introducing further contrail ice classes and may be implemented in the future.

The sedimentation parameterization and other components are as in the cirrus ice class.

The introduction of a second ice cloud class leads to a more complex behavior as both ice cloud classes are coupled and interact with each other in several ways. They are directly coupled via the collection process (see section 2.4) and the mass transfer of large contrail ice crystals as described above. Moreover, they interact indirectly via the competition for the available water vapor and possibly via dynamical changes through diabatic processes.

For a first illustration of our approach, Fig. 3.1 shows average size distributions (ASD) of a simulation with and without air traffic. Each ASD is a superposition

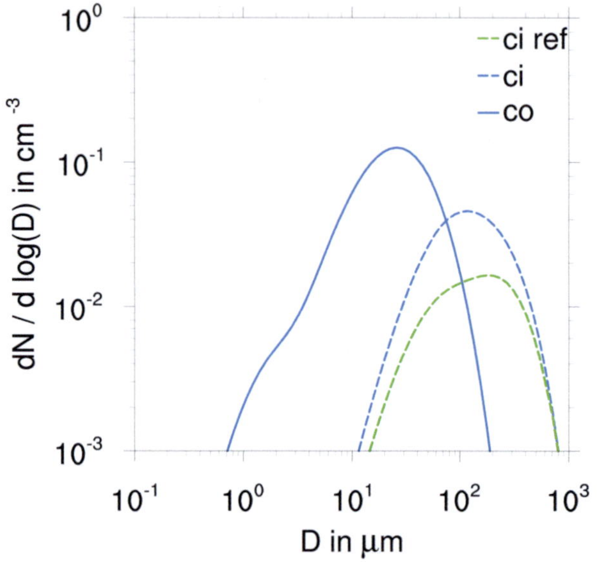

Figure 3.1: Ice crystal size distribution of the simulation with (blue) and without (green) aviation. For the simulation with aviation, separate ASDs of the contrail and cirrus ice class are shown (solid and dotted line). 12 UTC.

of local Γ-distributions. In this example, the contrails are at most 4 hours old. More details on the simulation setup follow in section 3.2.1. The solid blue line shows the contrail ice class which has a peak at sizes around 20 μm. A less pronounced maximum is located at about 2 μm. The contribution of aged contrails becomes apparent by comparing the two dashed lines. Those show the cirrus ice class ASDs of a simulation with air traffic (blue) and without air traffic (green). Apparently, the anthropogenic contribution is substantial, in particular in terms of total number. The ice crystals in aged contrails are on average smaller than in natural cirrus and the peak size of the SD is shifted to a smaller value.

Consequently, also in the radiation scheme, contrails are treated separately from the cirrus ice using the new contrail ice class. The applied radiation scheme is described in section 2.6. To include contrails and contrail cirrus in the radiative algorithm, we include a contrail ice cloud cover determined from the contrail ice class mass mixing ratio. Grid cells, where the ice mass mixing ratio exceeds 10^{-8} kg kg^{-1} are considered to be covered with contrails or contrail cirrus. The same threshold value is used in Seifert and Beheng (2006) for grid-scale natural ice clouds. As mentioned before, the aviation contribution to the natural cirrus ice class can only be determined by comparison with the reference simulation.

Within the radiative algorithm, optical properties of hydrometeors are calculated. These are the mass specific extinction coefficient, single scattering albedo, asymmetry factor and the forward peak of the phase function. For ice clouds, the parameterizations following Fu et al. (1998) and Key et al. (2002) are used. Because they are optimized for natural ice clouds, the scheme computes reliable values for effective radii r_e between 5 and 60 μm (Fu et al., 1998), whereas for ice crystal populations with radii smaller than 5 μm, the parameterization is not well defined. Ice crystals in young contrails often have effective radii smaller than 5 μm. To overcome this problem, we are using the parameterization of Fu et al. (1998) and Key et al. (2002), but we prescribe a lower limit of 5 μm for the calculation of the optical properties. For q_i, no limit is prescribed; instead, the simulated q_i is used to calculate the optical properties of the ice crystals. As the radiation scheme uses q_i and r_e for determining the radiative fluxes, implicitly fewer but larger crystals are assumed here. The extinction due to small ice crystals is expected to be larger than that for larger ones. Therefore, in our study, the radiative effect of young contrails may be underestimated.

The calculation of the contrail effective radii follows Fu et al. (1998). Here, the ice crystals are assumed to be randomly oriented, hexagonal columns.

3.1.2 Formation of contrails

For the description of contrails, the first step is to check whether the environmental conditions are favorable for the formation of contrails. Here, the Schmidt-Appleman-Criterion (Schumann, 1996) is used which defines a threshold temperature below which contrail formation occurs.
In the second step, the source term of contrail ice mass and number has to be calculated. The parameterization used to calculate those source terms is described in detail in Unterstrasser (2016).
Microphysical properties of aged contrails depend a lot more on the number of ice crystals than on the ice mass after the vortex phase (Unterstrasser and Gierens, 2010b). The initial ice mass is of minor importance, as the later growth of contrail ice crystals and the related ice mass evolution in a persistent contrail is mainly controlled by the ambient water vapor supply. On the other hand, the ice crystal number changes only slowly in a long-living contrail. Hence, its initial value determines the typical ice crystal sizes in the evolving contrail-cirrus (for a given environmentally controlled ice mass), which affects the radiative properties and the sedimentation-related life cycle.
Therefore, it is appropriate to explicitly prescribe an initial ice crystal number concentration instead of an initial ice mass.
The presented procedure is applied at each model time step and for each grid cell, given that both an aircraft is present and the Schmidt-Appleman-Criterion is simultaneously fulfilled.
The parameterization provides ice crystal numbers for contrails that are about 5 minutes old. As meteorological input parameters, it requires the temperature T at cruise altitude, the ambient relative humidity with respect to ice RH_{i} and the Brunt-Väisälä frequency N_{BV}. Furthermore, aircraft properties are characterized by the water vapor emission I_0, an 'emission' index for ice crystals EI_{iceno} and the wing span b_{span}.
We determine I_0 for medium fuel flow at cruise conditions as assumed in S. and N. (2014). Here, a simple parabolic fit for I_0 depending on the wing span

b_{span} is used (Unterstrasser, 2016). Information on the wing span is available in the flight track data (see section 3.1.3).

$$I_0 = c_1 \left[\frac{b_{\mathrm{span}}}{c_2}\right]^2; \quad c_1 = 0.02\ \mathrm{kg\ m^{-1}}; \quad c_2 = 80\ \mathrm{m} \tag{3.1}$$

In this study only the most common JET-A fuel is assumed to be used; therefore EI_{iceno} is set to 2.8×10^{14} $\mathrm{kg^{-1}}$ following Unterstrasser (2014).
The total number of ice crystals formed in the beginning, N_0, is calculated using the following equation:

$$N_0 = \frac{I_0}{EI_{\mathrm{H20}}} EI_{\mathrm{iceno}} \tag{3.2}$$

with water vapor emission index $EI_{\mathrm{H20}} = 1.25$. Note that EI_{iceno} is not reduced when the ambient temperature is only slightly below threshold temperature, even though in such situations fewer ice crystals would form (Kärcher et al., 2015).
The descending movement of the primary wake of an aircraft causes adiabatic heating within the plume. Due to this, sublimation and loss of ice crystals occurs, even in a supersaturated environment (e.g. Unterstrasser, 2016). As the spatial and temporal resolution of the model is too coarse to simulate these processes, the fraction of ice crystals surviving the vortex phase is parameterized by a loss factor λ_{Ns}. Details can be found in Unterstrasser (2016). The total number of surviving ice crystals per flight path N_{s} is calculated with:

$$N_{\mathrm{s}} = N_0 \lambda_{\mathrm{Ns}} \tag{3.3}$$

For the initial ice mass produced, the water vapor emission I_0 is used. The values for the produced ice mass and ice crystal number per flight distance are distributed equally on all grid cells, the aircraft passes within a time step:

$$n_{\text{init}} = \frac{N_s d}{V_{\text{cell}}}; \quad q_{\text{init}} = \frac{I_0 d}{V_{\text{cell}}} \tag{3.4}$$

Here, V_{cell} denotes the volume of the grid cells, d is the flight distance within a grid box.

In global models, contrail parameterizations usually contain further prognostic equations that describe in some way the bulk contrail geometry (e.g. fractional cloud coverage or even some measure of contrail depth). In our approach, we assume that contrail ice crystals always populate the whole grid box as the horizontal scale is much smaller than in GCMs. In the vertical direction, it is assumed that ice crystals are distributed over the whole grid layer. Close to the ground, the vertical grid size is about 10 m and increases to 300 m at the tropopause. In supersaturated conditions, contrail depth varies between 100 m and 500 m (mostly depending on aircraft type and stratification) and is similar to the depth of the grid layer. In the horizontal plane, the simplifications could have a larger effect. If few flight routes transect a grid box and the segments are in total $d_{\text{AT}} = 10$ km long then this implicitly results in an hypothetical initial contrail width of $\frac{2.8 \times 2.8}{10}$ km ≈ 800 m. This is larger than what is typically observed for 5 minute old contrails and better fits to 15 minute contrails (Freudenthaler et al., 1995). Hence, disregarding fractional coverage smears out the initialized contrails to some extent.

3.1.3 Aircraft movement data

In contrast to most of the previously mentioned global modeling studies, this study uses a new and recently derived data set. Rather than statistical calculations for globally averaged fuel consumption, or radar data, the basic data consist of traffic waypoint information over a limited area recorded from ADS-B[1] transponders on the plane (flightradar24.com, 2015). The ADS-B data is obtained mainly from flightradar24.com (2015). The DLR holds a historic data file that can be purchased from flightradar24.com (2015). This data is cleaned and combined with the input of the Official Airline Guide Database that is also hold by DLR.

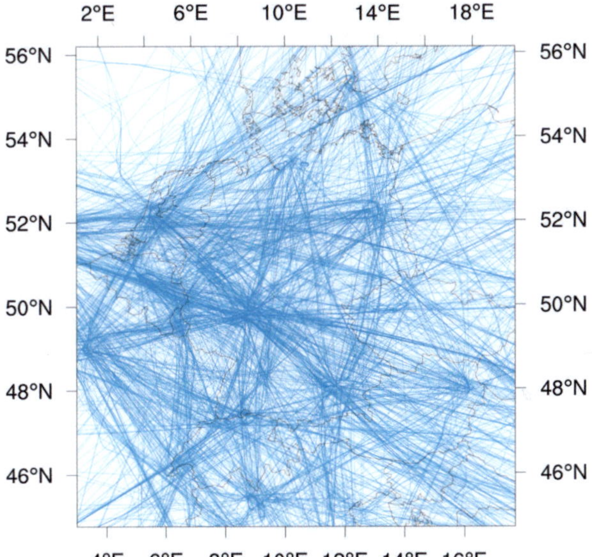

Figure 3.2: Flight trajectories for the simulated domain derived from ADS-B data (08 UTC - 16 UTC 3 December 2013). Each trajectory is plotted as a faint dashed line. Seemingly solid or thick lines are indicative of multiple overlaid trajectories.

From this information, a dataset is compiled containing spatially and temporally resolved information on geographical position, height, current velocity and type of aircrafts. For reasons of efficiency, the information on geographical position is interpolated to fit onto the grid of the model. As a

[1] Automatic Dependent Surveillance - Broadcast

proxy for the aircraft emission parameters (see section 3.1.2), the wing span b_{span} is used. The data set contains eight hours of air traffic, beginning at 08 UTC 3 December 2013 and ending at 16 UTC 3 December 2013. The trajectories of all flights during this period are displayed in Fig. 3.2.

3.2 Case study

To test the methods described in the previous sections, a case study was performed. For this purpose, a situation over Germany with a high density of contrails in an otherwise cloud-free environment was chosen as the simulation period.

3.2.1 Model setup

On 3 December 2013, the meteorological conditions over Central Europe were favorable for the formation of contrails. Additionally, the natural high-cloud coverage was relatively low, thus allowing for the identification of contrails on satellite images.
For the case study, two simulations were conducted, both running for 24 hours, starting on 3 December 2013, 00 UTC. They use a horizontal 2.8×2.8 km grid and 60 vertical levels, resulting in a mean distance between the model layers of 300 m in the upper troposphere and 15 m in the lowest model layer. For boundary data, hourly COSMO re-analyses were used. In the reference run, air traffic is turned off and the cirrus ice class is the only ice cloud class to be active. For the run with air traffic, which we call aviation simulation, the previously explained configuration with two ice cloud classes is employed. Air traffic is switched on at 08 UTC and the two simulations evolve identically up to this point. Practically, a spin up phase shorter than 8 h could have been used, but from an operational point of view it was simpler to start the simulations at 00 UTC. As the data set of flight trajectories contains no information about air traffic after 16 UTC, no new contrails form after this time.

3.2.2 Simulated contrail properties

The contrail treatment in the microphysics scheme is designed such that contrail-induced changes occur both in the newly introduced contrail ice class and in the existing cirrus ice class. The contribution of young contrails can be directly assessed by evaluating the contrail ice class. The contribution of aged contrails is found by comparing the cirrus ice class of the aviation simulation and of the reference simulation.

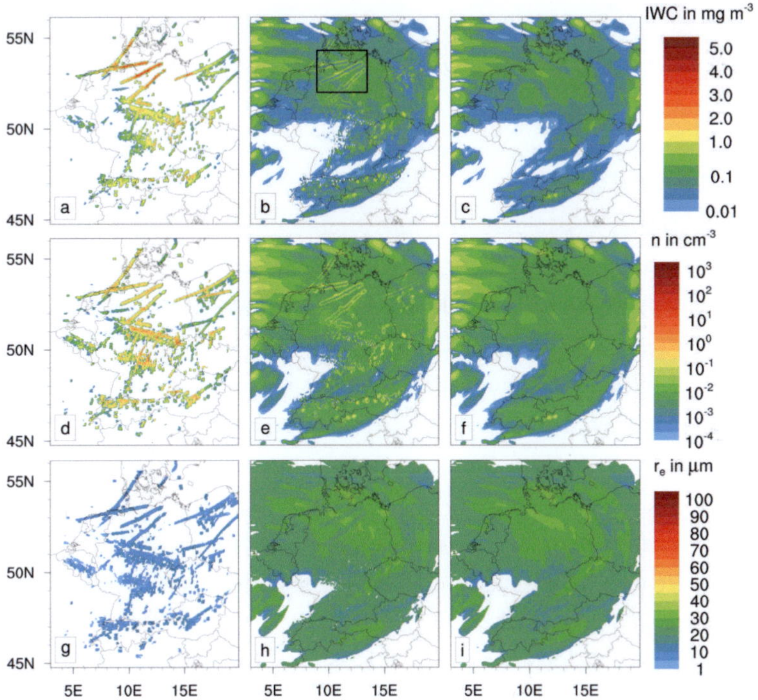

Figure 3.3: Various ice cloud properties are depicted for 10 UTC at an altitude of 11900 m (top row: ice water content; center row: ice crystal number concentration; bottom row: effective radii). Depicted are the simulation with aviation (left column: contrail ice class; middle column: cirrus ice class) and without aviation (right column).

Figure 3.3 shows ice cloud properties over Central Europe in the model layer centered at $z = 11900$ m, at 10 UTC. The various rows (from top to bottom) show the ice water content IWC, the ice crystal number concentration n, and the effective radius r_e. The left column shows the contrail ice class of the aviation simulation, whereas the middle and right columns show the cirrus ice class of the aviation and the reference simulation. At this time, contrails are at most two hours old and mostly consist of numerous very small ice crystals.
Independent of the considered quantity, the contrail ice class features mostly line-shaped structures. The IWC reaches values up to 5 mg m^{-3}, which is larger than in the simulated natural cirrus. This hints at an accumulation of emitted water vapor additional to the ambient supersaturation. Furthermore, the absolute humidity at the heights considered is relatively low. Therefore, the IWC of natural cirrus is small and cirrus clouds are very thin and almost invisible. The ice crystal number concentrations often lie between 1 cm^{-3} and 100 cm^{-3} and can exceed those in natural cirrus by a factor of up to 1000. Consequently, the ice crystal effective radii in contrails are much smaller and lie below 10 μm.
The middle panel in Figure 3.4 shows that the most dense contrails over Northern Germany already leave a mark in the cirrus ice class (see black box in panel b)). Notably, each line-shaped structure in the left panel corresponds to a pair of lines in the middle panel. This indicates growth of ice crystals particularly at the margins where the contrail ice mass is soon transferred to the cirrus ice class. Consistently, large-eddy simulation studies (Unterstrasser and Gierens, 2010a; Lewellen et al., 2014) and in-situ measurements (Petzold et al., 1997; Heymsfield et al., 1998) indicate the strongest growth at the edges of a contrail. A closer inspection (not shown) reveals that each line-shaped structure in the aforementioned box consists of several contrails. Those were produced by several aircraft that fly along the same route with short time separations. Nevertheless, the IWC, n and r_e-values of the aged contrails and the surrounding natural cirrus shown in Fig. 3.3 are similar.

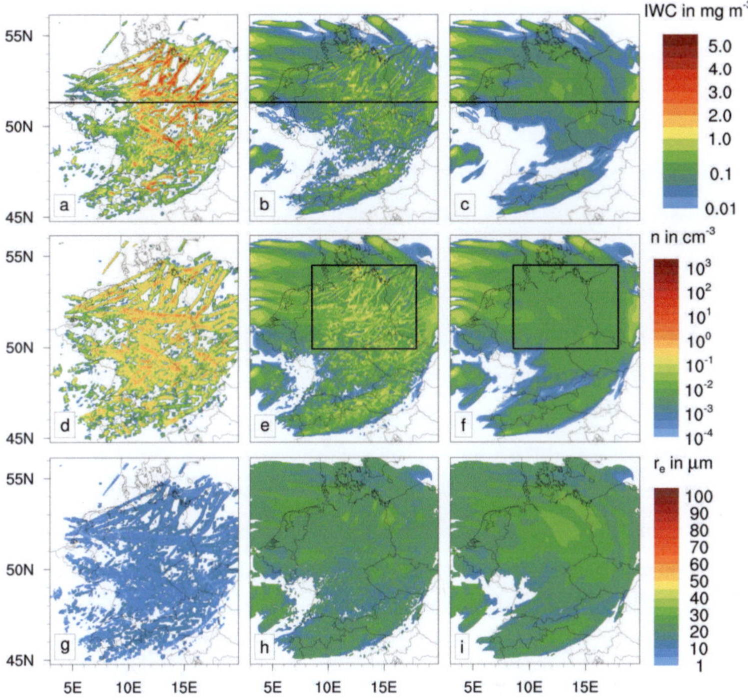

Figure 3.4: Analogous to Fig. 3.3, now for 12 UTC. The black horizontal lines indicate the location of the curtain displayed in Figs. 3.5 and 3.7; black boxes are discussed in the text.

Figure 3.4 shows the situation at 12 UTC, analogous to Fig. 3.3 for 10 UTC. As time progresses, existing contrails continue to grow and further contrails are generated. Four hours after air traffic was activated in the model, a major part of the model layer is filled with contrails (left panel). Line-shaped patterns are still identifiable in some places, particularly over Northern Germany. South of 52°N many contrails overlap and represent a huge contrail cluster. Based on their visual appearance, no clear distinction from natural cirrus would be possible. Those contrail clusters still feature high IWC, high n, and low r_e values comparable to those two hours earlier and distinct to the surrounding natural cirrus. Moreover, we find lots of aged contrails over Germany (black box in middle panel), a region that is basically cirrus-free if there is no air traffic (black box in right panel). Here, local enhancements in both IWC and n occur in the cirrus ice class, where aged contrails are transferred to (Fig. 3.4e), compared to the reference simulation (Fig. 3.4f).

The results indicate that, in observations, microphysical criteria may help to separate at least young contrails from natural cirrus. In general, contrail fall streaks and aged contrails cannot be identified as such once their linear shape is lost or masked. Unterstrasser et al. (2017b) shows that natural cirrus that forms in high-updraft scenarios can have ice crystal numbers similar to those of young contrails, which renders even the separation between young contrails and natural cirrus impossible. Moreover, they show that contrails that become embedded in natural cirrus have large volumes where ice crystals of both origins co-exist. Hence, it is no longer meaningful to try to draw a strict separation line between natural cirrus and the anthropogenic cloud contribution.

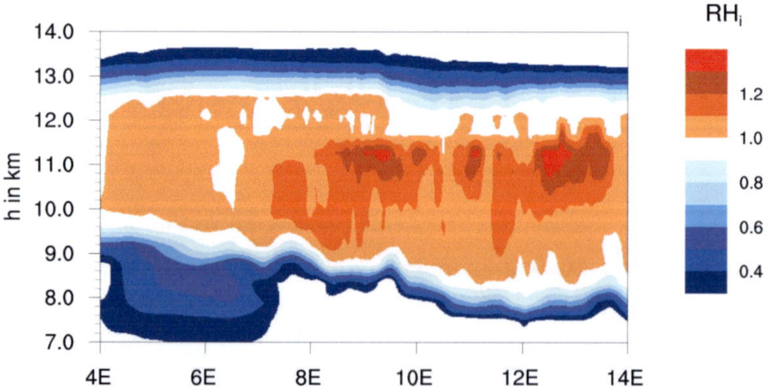

Figure 3.5: Vertical cross section of relative humidity RH_i at 12 UTC along the black line in Fig. 3.4.

Next, we analyze vertical distributions along the black line depicted in Fig. 3.4. Fig. 3.5 displays the relative humidity RH_i and reveals a remarkably thick layer with strong supersaturation (maximum value: 1.34) that extends over the complete east-west extent of the model domain. The layer depth increases from 2.5 km in the west to more than 4 km in the east. These are generally very favorable conditions for the persistence and spreading of contrails.

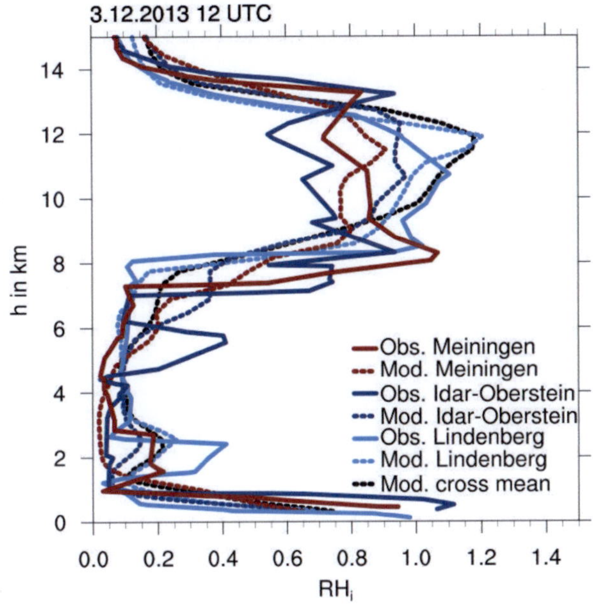

Figure 3.6: Radio soundings for several locations (solid lines) (UWYO, 2018) and corresponding profiles from aviation simulation (dotted lines) of relative humidity RH_i. The black dotted line are mean values along the black line in Fig. 3.4.

The fidelity of such high values of RH_i is corroborated by a comparison with observations. Here, vertical profiles obtained from radio soundings (UWYO, 2018) and simulated data evaluated at radiosonde stations are depicted in Fig. 3.6. In general, high values of RH_i are observed by radio soundings, even supersaturation occurs over Lindenberg and Meiningen.

The model clearly overestimates RH_i for Idar-Oberstein, however Meiningen and Lindenberg agree quite well. Additionally, the black curve shows the mean vertical profile of the cross section displayed in Fig. 3.5. From this it becomes apparent that the relative humidity in the displayed cross section is remarkably high compared to the radiosonde locations. The large layer of supersaturation is caused by lifting and radiative cooling. It can persist, as natural cirrus clouds are located mostly below this layer (Fig. 3.7b). Also, the cirrus clouds present in the layer are too thin, i. e. occurring number concentrations are too low (Fig. 3.7e) to effectively reduce supersaturation.

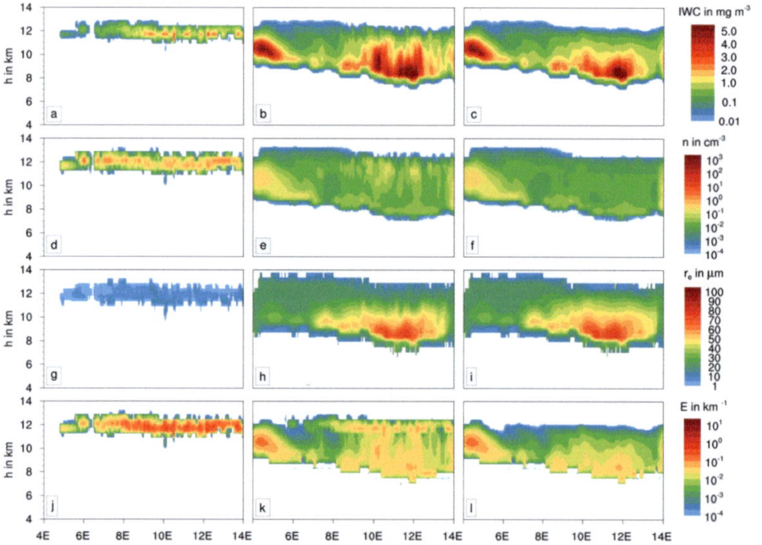

Figure 3.7: Vertical cross section of various contrail properties at 12 UTC along the black line in Fig. 3.4: ice water content, ice crystal number concentration, effective radius, and extinction coefficient (from top to bottom). Depicted are the simulation with aviation (left column: contrail ice class; middle column: cirrus ice class) and without aviation (right column).

Figure 3.7 shows the same ice cloud properties as Fig. 3.4 and additionally the extinction coefficient E at a wavelength of 1.115 μm. Again, the left column shows the contrail ice class of the aviation simulation, whereas the middle and right column show the cirrus ice class of the aviation and the reference simulation.

In the reference simulation, natural ice is present over the entire supersaturated area. The number concentrations are mostly small, leading to optically thin cirrus clouds with extinction coefficients hardly exceeding 10^{-2} to 10^{-1} km^{-1}. Rather large ice crystals are present throughout the supersaturated area, with the largest values of r_e found in the lower part of the cirrus around 10 to 14° E.

In the aviation simulation, it becomes obvious from the left column plots that contrails form only at altitudes between 11 km and 13 km. This is caused by the absence of air traffic below this layer. As already seen in Fig. 3.4a;b, the IWC is comparable to that of natural cirrus, whereas number concentrations reach much higher values.

The effective radii in the aviation simulation are typically one order of magnitude smaller than in natural cirrus clouds and do not exceed 10 μm. These results are in agreement with large-eddy studies (Unterstrasser and Gierens, 2010a; Lewellen et al., 2014) and in-situ observations (Poellot et al., 1999). The numerous small crystals lead to high values for the extinction coefficient (up to 2 km^{-1}). Therefore, contrail ice is of great importance for the radiation budget.

In Fig. 3.8, the relative occurrence of ice crystal number concentrations and temperature for the cross section shown in Fig. 3.7 is depicted. The relative occurrence is normalized with the sum over all values. Both, reference simulation (Fig. 3.8a) and aviation simulation (Fig. 3.8b) are similar for higher temperatures (i. e. lower heights) up to 220 K.

For lower temperatures, high number concentrations up to 7 cm^{-3} occur in the aviation simulation, whereas number concentrations clearly decrease strongly with temperature in the reference simulation. Here, a rough comparison to measurement data can be made. In Voigt et al. (2017), mid-latitude cirrus clouds and contrails where probed in-situ during an aircraft measurement campaign. Comparing their Fig. 6(b) to Fig. 3.8, a similar increase in n below temperatures of about 220 K is found. Therefore, most likely, the high values

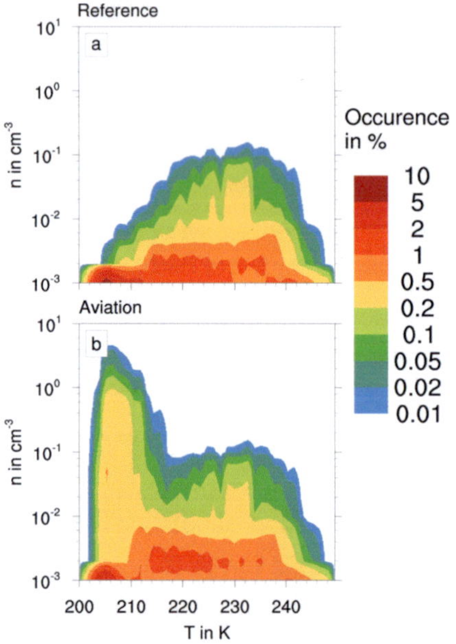

Figure 3.8: Relative occurrence of ice crystal number concentration versus temperature for natural cirrus, contrail and contrail cirrus at 12 UTC for the cross section along the black line in Fig. 3.4; a) reference simulation; b) aviation simulation.

occurring in the aviation simulation and not in the reference simulation, are due to aviation induced clouds.

In the aviation simulation, changes in the natural ice class can be found mainly at heights where contrails form and slightly below of it. The enhancement of ice number concentrations occurs mostly at flight levels, whereas an increase in IWC is found below. During the initialization, contrail ice crystals are vertically distributed over the whole grid layer and this indirectly accounts for the initial wake vortex induced vertical expansion of a contrail. Within our simulation, the vertical structure of the contrails is determined only due to the gravitational settling of the larger ice crystals. During this process, ice crystals number concentrations tend to decrease. In contrast, only a slight increase in r_e is found. In areas where contrail ice enters the cirrus ice class, a large increase in extinction coefficient occurs. Values of the extinction coefficient are comparable to those of the contrail ice class and can be as high as 1 km^{-1}.

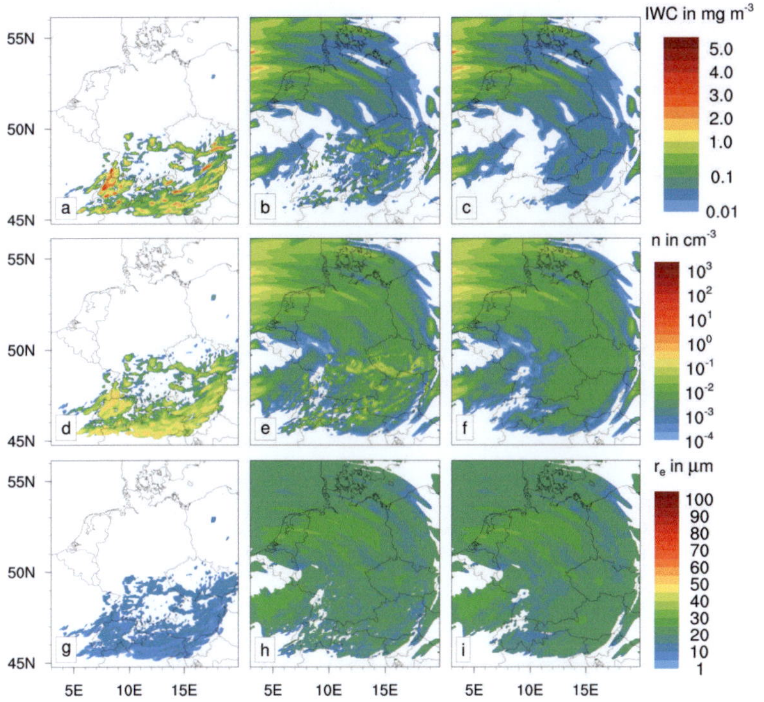

Figure 3.9: Analogous to Fig. 3.3, now for 20 UTC

After 16 UTC, no new contrails are initialized in our simulation. Four hours after the end of new contrail formation, the remaining contrail ice has been advected to the southern part of domain (Fig. 3.9). Local patterns of increased number concentrations in the cirrus ice class are now limited to those regions, where contrail ice is still present. The line shaped structures in the contrail ice class vanish, but relatively small values for r_e are still found throughout the domain.

In the northern part of domain, the cirrus ice class is again undisturbed by aviation. Here, no remarkable differences to the reference simulation can be seen.

3.2.3 Comparison with satellite observation

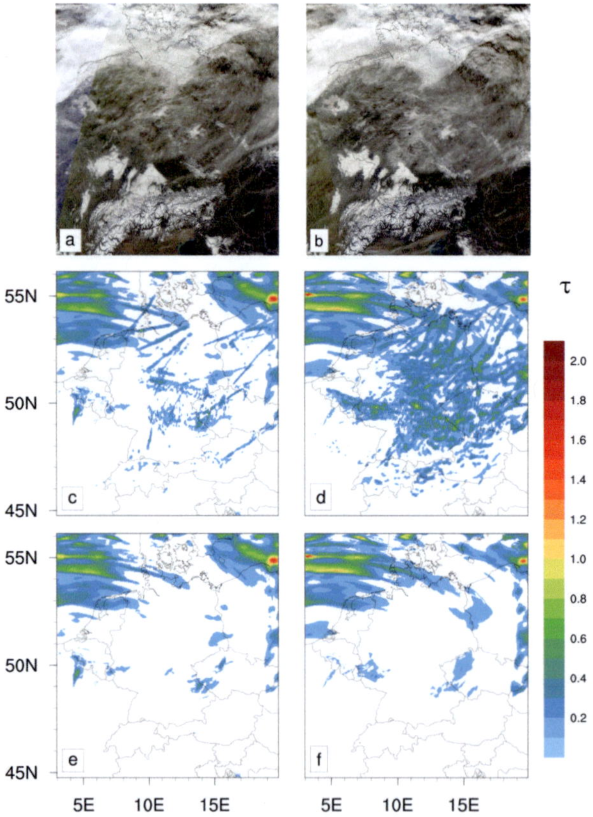

Figure 3.10: Top row: satellite image (MODIS True Color - Corrected Reflectance) (NASA/GSFC/ESDIS, 2018); center row: optical depth at 1.115 μm of all ice clouds for the aviation simulation; bottom row: optical depth of all ice clouds for the reference simulation; left column: 10 UTC; right column: 12 UTC.

In the following, satellite images (created with Global Imagery Browse Services (GIBS) NASA/GSFC/ESDIS, 2018) are shown in Fig. 3.10 for a qualitative assessment of the simulations. The panels a and b show the "MODIS Terra Corrected Reflectance True Color" at 10 UTC and 12 UTC for the simulated day, respectively, both with a resolution of 250 m. The "True Color" composition consists of MODIS bands 1, 4 and 3 (NASA/GSFC/ESDIS, 2018). Beside a cloud bank over the North and Baltic Seas and fog over Southern and Western Germany, a considerable number of line-shaped contrails and diffuse cirrus clouds are present across both satellite images. Main contrail clusters are found over Central Germany for both situations. Contrails can also be identified over the Netherlands and Belgium, over the Czech Republic, and south of the Alps. At 12 UTC, the high-level cloud cover has increased and pervaded compared to 10 UTC.
Comparing Fig. 3.10a and Fig. 3.10e, obviously, the reference simulation underestimates the coverage of high clouds in the center of the domain, which, in large parts, consists of contrails and contrail cirrus Fig. 3.10c. Although a number of the patterns of high-level cloud cover detected by MODIS at 10 UTC can be identified in the aviation simulation, the amount of cloud cover seems to be slightly underestimated also in the aviation simulation at 10 UTC. This discrepancy is probably due to the fact that air traffic was switched on at 08 UTC and earlier flight movements are disregarded in our simulation. Hence, the simulation evaluation at 10 UTC neglects all contrails older than 2 hours. The comparison at 12 UTC is more favorable than the one at 10 UCT. Observations match a lot better with the aviation simulation than with the reference simulation, as in the aviation simulation, a much larger fraction of contrails could form since 10 UTC. Mostly over the center of the domain, areas with values of τ between 0.2 to 0.6 and peaks up to 1.0 are simulated. This is in good agreement with the very high values of the extinction coefficient of contrails compared to those of natural cirrus (Fig. 3.7j, Fig. 3.7k, Fig. 3.7l). The line shaped patterns stem from the contrail ice class and the more patchy structures from aged contrails which have been transferred to the

cirrus ice class. The case study for this particular day demonstrates that the inclusion of aircraft effects in a regional weather forecast model improves the representation of high level clouds.

3.2.4 Contrail impact on surface radiative fluxes and photovoltaic power production

Figure 3.11 shows the changes in the incoming shortwave radiation (SW) at two points in time at the surface. Clearly, additional ice crystals caused by air traffic reduce the incoming SW radiation. The line-shaped structure of young contrails at 10 UTC (Fig. 3.3a) is reflected in similar patterns of reduced SW radiation (Fig. 3.11a). As discussed earlier, the contrail ice class features numerous very small ice crystals with a high extinction coefficient. In the case investigated, this leads to a reduction in incoming SW radiation of 1 to 15 Wm^{-2}. As previously mentioned, our parameterizations are not able to calculate the optical properties of ice crystals with effective radii below the threshold of 5 μm. Rather, crystals with an effective radius of less than 5 μm are assumed to have optical properties of crystals as large as 5 μm. Therefore, the shading effect of the contrail ice crystals might be underestimated in our calculations.

A strong spatial increase in the shading effect is found at 12 UTC (Fig. 3.11b). Here, the contrail coverage reaches its maximum. Still partly line shaped contrails are found with a similar reducing impact as two hours before. Additionally, clusters of contrail ice transformed into the cirrus ice class have evolved, particularly over the center of the domain (Fig. 3.4b). The ice crystals herein are relatively small, but larger than those present in the contrail ice class. Therefore, they inhibit less SW radiation from reaching the ground than the ice crystals in the contrail ice class. Consequently, the reduction on SW radiation here is smaller, but still reaches 1 to 10 Wm^{-2}.

Especially in the north of the domain, small areas with the difference in incoming SW radiation attaining large negative values adjacent to large positive values are found. They occur when subtracting e. g. fields of radiative fluxes of the aviation simulation from those of the reference simulation. The introduction of contrail ice acts as source of disturbance for various processes like convection or turbulence. Those features are to be classified as noise as they do not influence the overall situation.

Non-negligible areas with an increase in SW radiation occur, e. g. at 12 UTC over the south-eastern part of the domain (Fig. 3.11b). Besides only reducing the direct incoming SW radiation, contrail ice crystals also enlarge the flux of diffuse SW radiation (see below). On occasion, as the simulated contrails are still rather optically thin, this effect may be larger than the reduction of direct radiation.

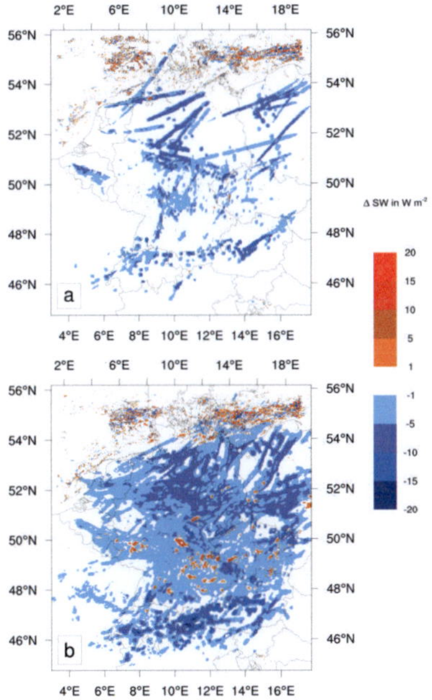

Figure 3.11: Difference in shortwave incoming radiation for aviation simulation and reference simulation for 3 December 2013 at the surface, a) 10 UTC, b) 12 UTC.

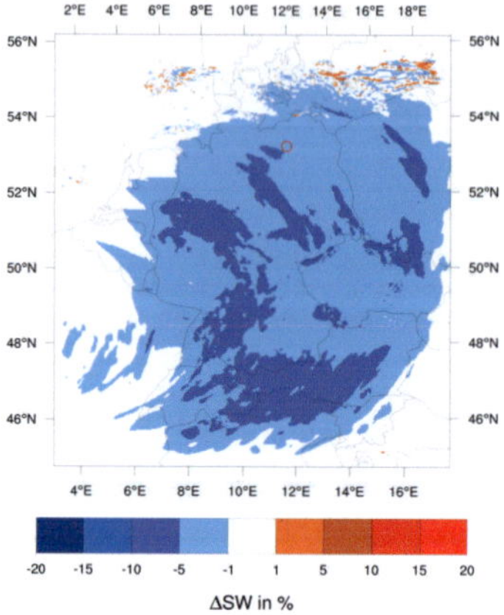

Figure 3.12: Average changes in the amount of incoming diffuse and direct solar radiation during daylight hours of 3 December 2013 at the surface (aviation simulation minus reference simulation). The red circle (53°N, 12°E) marks the location evaluated in Fig. 3.13.

The average change of incoming direct and diffuse SW for 3 December 2013 is shown in Fig. 3.12. Also here, the small scale fluctuating values in the north are most likely noise.

The thin veil of contrail cirrus that spreads over most of the domain causes an average decrease of incoming direct and diffuse SW radiation of 1 to 5 %. The large and persistent contrail cluster over the northern and eastern part of the domain inhibits on average 5 to 10 % of SW radiation from reaching the ground during approximately eight hours of daylight. The reduction is strongest over the south and the west of the domain. Here, contrail ice is present for the longest time, as seen in Fig. 3.9.

Fig. 3.13 shows the temporal evolution of incoming SW radiation and normalized PV power. Panels a to d represent mean values for the entire simulation domain, panel e additionally represents mean values for an area of approximately 50 km × 50 km centered at a spot in the north-eastern part of the simulated domain, indicated by the red circle (53°N, 12°E) in Fig. 3.12. This is the location of one of the largest solar parks in Germany (Solarpark Brandenburg-Briest). Also several other major solar parks are located around this area. For enhanced clearness, values for the sensitivity study (see section 3.2.5) are depicted only as difference to the reference simulation. The differences between the aviation simulation and the reference simulation are in general more pronounced for the selected location that on average over the entire simulation domain.

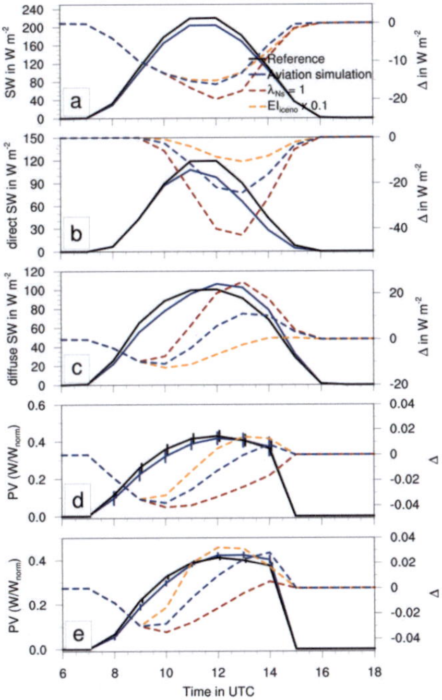

Figure 3.13: Temporal evolution for 3 December 2013 of incoming shortwave radiation: reference (black); aviation simulation as before (blue); "bio-fuel" scenario with $EI_{\text{iceno}} \times 0.1$ (orange), omission of crystal loss during contrail vortex phase with $\lambda_{\text{Ns}} = 1$ (red). Dashed lines (corresponding to right y-Axis) are difference to reference simulation. a) total SW; b) direct SW; c) diffuse SW; d),e) normalized PV-power. a) - d) over entire domain; e) mean for the location marked with the red circle in Fig. 3.12. Error bars represent mean values +/- standard deviation.

During the whole time of daylight, contrails and contrail cirrus reduce up to 15 Wm^{-2} (about 7 %) of the total incoming SW radiation in the entire domain (Fig. 3.13a). The effect is largest during noon and ceases during the day. This corresponds to the size of contrail ice crystals. As in our simulation, contrails start to form at 08 UTC, the average contrail ice crystal size grows during the day. Accordingly, contrail ice effective radii also increase with time and lead to smaller values of the extinction coefficient.

Separating the total SW radiation into its direct (Fig. 3.13b) and diffuse (Fig. 3.13c) fraction, it is clear that especially the direct incoming SW radiation is strongly reduced by more than 20 Wm^{-2} due to the presence of additional ice crystals in the atmosphere. In contrast, the diffuse incoming SW radiation is increased by up to 10 Wm^{-2}. Enhanced scattering of SW radiation caused by the contrail ice crystals increases the diffuse SW radiation at the ground.

Notably, the peak in reduction of diffuse SW radiation occurs in the afternoon, around 13 UTC. Between 08 UTC and about 11 UTC, the amount of diffuse SW radiation reaching the ground is larger in the aviation simulation. During this time, as mentioned before, contrail ice crystals are smallest on average and forward scattering is less pronounced than for larger crystals, whereas contrail ice crystals grow on average during the day resulting in enhanced forward scattering.

In the aviation simulation, young and aged contrails generally reduce the incoming SW radiation at the surface. This effect is currently neglected in operational weather forecast models. However, this effect is of relevance for the production of solar energy. The temporal evolution of the normalized PV power is depicted in Fig. 3.13d) and Fig. 3.13e. The normalized PV is calculated using the open source PV modeling environment PV_LIB for python (Andrews et al., 2014). For a specific combination of a PV module and a PV inverter combination, a nominal power and a reasonable tilt is assumed. These technical specifications are taken from Rieger et al. (2017). They assume panels consisting of a south-oriented PV module with a nominal power of 220W and a size of 1.7m^2. Compared to the reference simulation, the normalized PV power is decreased in the aviation simulation most of the day. The largest losses of up to 10 % occur in the morning and diminish during the day, even an increase occurs for the selected location in the late afternoon (Fig. 3.13e). The normalized PV power is somewhat more strongly reduced than the total SW radiation. For production of PV power, the incoming direct radiation is of greater importance than the diffuse; of the two, the direct experiences the larger reduction from contrails and contrail cirrus.

The error bars in Fig. 3.13d and Fig. 3.13e represent the mean values +/- the standard deviation with respect to the entire simulation domain and the area of 50 km $\times$ 50 km around the selected location, respectively. The standard deviation in Fig. 3.13d reflects the large-scale variability of the impact of contrails and contrail cirrus on the incoming SW radiation, whereas Fig. 3.13e illustrates the small-scale variability.

Compared to the selected location, standard deviations are larger for the mean of the domain. Obviously, clouds modify the amount of SW radiation reaching the ground in a non-uniform manner. The magnitudes of the standard deviations for the aviation simulation are about the same magnitude like the ones for the reference simulation. Apparently, the impact of contrails and contrail cirrus on incoming SW radiation is as variable as the impact of natural clouds. E. g., even at 12 UTC, when the impact of contrails and contrail cirrus is largest, confined areas exist, which are unaffected by contrails and contrail cirrus (Fig. 3.11b).

However, the small-scale variability of the impact of contrails and contrail cirrus on SW radiation is rather small, reflected by much smaller standard deviations in Fig. 3.13e. One can therefore deduce that the exact location of contrails or contrail cirrus is not crucial for the strength of the impact on SW radiation.

3.2.5 Sensitivity to initial ice crystal number and early contrail ice crystal loss

In this last section, we briefly examine two sensitivities of our model setup. For the first, we reduce the emission index for ice crystals EI_{iceno} by a factor of 10 (orange lines in Fig. 3.13). This scenario explores lower engine soot emissions caused by either improved engine combustion technologies or fuel composition changes from, e. g., biofuel adoption (Moore et al., 2017). Due to this, the initial number concentration of contrail ice crystals is reduced, thus fewer but larger ice crystals are formed (Unterstrasser, 2014). In the simulation with reduced EI_{iceno}, the reduction of total incoming SW radiation is only slightly weaker than for the simulation assuming standard fuel (Fig. 3.13a). As ice crystals are slightly larger, their extinction coefficient is lower and the reduction of direct SW radiation is smaller than for the standard setup (Fig. 3.13b). Also the increase in incoming diffuse SW radiation is less strong compared to the standard setup (Fig. 3.13c). Consequently, the reduction in normalized PV power is also less strong than for the standard setup, even an enhancement occurs during afternoon (Fig. 3.13d, Fig. 3.13e).

Second, we set the surviving fraction of ice crystals λ_{Ns} in the parameterization of the initial ice crystal number to 1. This deliberately neglects the effects of crystal loss during the contrail vortex phase, as parametrized by Unterstrasser (2016). The influence of this parameter is large. During daytime, a reduction in total incoming SW radiation of up to 15 % is simulated (Fig. 3.13a). Both the reduction in direct SW radiation as well as the increase in diffuse SW radiation are much more pronounced for this case (Fig. 3.13b, Fig. 3.13c). Especially the reduction in direct SW radiation causes a strong reduction in production of PV power. Here, losses of nearly 15 % occur at about 10 UTC. Also concerning the temporal evolution, the reduction lasts much longer compared to the aviation simulation.

When no early crystal loss is parametrized in the model, initial ice crystal number concentrations may be much higher than usual. As the initial IWC

remains the same, the new crystals are smaller and thus, the simulated contrails are optically thicker. The much stronger reduction in incoming SW radiation demonstrates that the early ice crystal number loss is non-negligible and an important aspect of contrail evolution as it has a long-lasting impact on contrail-cirrus radiative properties.

3.3 Summary

In this study, the regional atmospheric model COSMO-ART coupled with a two-moment microphysical scheme and a diagnostic treatment of radiation was extended by a parameterization describing contrails and the related physical processes.

Methods for a separate but consistent treatment of contrail ice were implemented to satisfy the special requirements describing the microphysics in young contrails and the transition phase to contrail cirrus. As the contrail-related parts of the model are entirely online coupled, feedback processes on natural clouds and the environment can be studied. In using such an approach, the presented study is complementary to former ones applying either LES models or GCMs. Compared to using a GCM, the microphysical processes within this study are described explicitly and in more detail; compared to LES, a larger number of contrails can be investigated. Furthermore, the simulations conducted with COSMO-ART can easily be validated with satellite data.

Performing a case study for a single winter day over Central Europe, it was shown how microphysical properties such as ice water content, ice crystal number concentration and the mean ice crystal radius of ice crystals in contrails change over time and depend on the meteorological conditions. The ice water content in young contrails is comparable to that in thin cirrus clouds ranging from 0.2 to 5.0 mg m^{-3}, but with considerably higher ice crystal number concentrations between 1 cm^{-3} and 100 cm^{-3} and effective radii below 10 μm. The numerous small ice crystals produce high values for the extinction coefficient and thus also for the optical depth.

The transition of contrail ice into the cirrus ice class causes increasing number concentrations. Here, effective radius of the ice crystals from the contrail ice class grows to an extent comparable to that of natural cirrus. Because of the still relatively high number concentrations, contrail cirrus still features high values for both the extinction coefficient and the optical depth.

Qualitative comparison with satellite data shows good agreement and proves advantages of considering contrails and contrail cirrus in a regional weather forecast model.

Contrail cirrus tends to cause changes in the microphysical appearance of high-level cloud coverage to a remarkable extent, which in turn influences the radiative effect in these regions.

In addition, the impact of contrails and contrail cirrus on shortwave radiation and the production of PV power were quantified. Although the case study was performed for 3 December 2013, when solar zenith angles are low and the length of days is short, a strong influence of contrails is still simulated. They inhibit up to 5 to 10 % of shortwave radiation from reaching the ground at noon. This results in a loss of PV power production of up to 10 %.

Moreover, it was demonstrated that ice crystal loss in young contrails is an important process which can significantly change the contrail-cirrus properties on a regional scale. So far the importance of crystal loss was only shown for single contrails with a LES model (Unterstrasser, 2014).

This study is the first approach to simulate contrails and contrail cirrus using a numerical weather prediction model with high spatial and temporal resolution. Subsequently, the presented method can serve as a basis for improving the predictability of the solar radiation in regional weather forecasting by taking into account contrails and contrail cirrus.

4 A process study on thinning of Arctic winter cirrus clouds with high-resolved ICON-ART simulations

This study aims at getting detailed insight into the microphysical processes affected by realistic seeding scenarios using the temporal and spatial high resolution modeling system ICON-ART coupled with a sophisticated two moment microphysical scheme (Seifert and Beheng, 2006).
In large parts, this chapter is based on a manuscript given the title *A process study on thinning of Arctic winter cirrus clouds with high-resolved ICON-ART simulations*. Currently (October 2018) the manuscript is under review in *Journal of Geophysical Research - Atmospheres*. The author of this thesis is also the lead author of the manuscript. However, valuable contributions to the manuscript of Ulrich Blahak, Florian Haenel, Christoph Kottmeier, Thomas Leisner, Harel Muskatel, Trude Storelvmo, and Bernhard Vogel are gracefully acknowledged.
The effects of seeding aerosol with nucleating properties like mineral dust injected in various scenarios into the atmosphere are explored. Furthermore, due to the high resolution, the impact of cirrus seeding on lower-lying mixed-phase clouds can be investigated, as the effect of seeding here has been proposed to outweigh much of the desired outcome (Gasparini et al., 2017).
As several studies find seeding being most effective in polar winter conditions, the focus is on a period of the Arctic winter 2015 / 2016. During this time,

the combined POLSTRACC[1]/GW-LCYCLE II[2]/SALSA[3] campaign, in the following abbreviated by 'PGS', took place and is one of the rare hibernal Arctic airborne field campaigns in the recent past.
The chapter is structured as follows. In section 4.1, the decisive parts of the model setup are briefly described. A detailed comparison with available in-situ and satellite data in section 4.2 allows gaining insight into the formation processes and microphysical properties of cirrus clouds during the Arctic winter. In addition, the impact of atmospheric background aerosol involved in both homogeneous and heterogeneous nucleation, as well as different assumptions the parameterization of the sub-grid scale variance of vertical velocity based on turbulent kinetic energy (TKE) (Lohmann et al., 1999) are investigated.
In section 4.3, the impact of seeding aerosol on both cirrus and mixed-phase clouds and the consequential effects on LW radiation are investigated. Furthermore, assumptions made in former studies regarding the properties of seeding aerosol, like the optimal concentrations, size of the aerosol particles , and possible strategies of deployment, are reviewed.

4.1 Model setup

In this study, ICON-ART is operated in a so-called limited area mode. Unlike global modeling systems, only a certain area is selected as simulation domain, and lateral boundary conditions for meteorological variables are updated hourly. This allows the conduction of a large number of computational relatively inexpensive simulations.
For the presented case study, the focus is on an area covering Greenland and parts of the Arctic Ocean including the northern parts of Scandinavia and Canada (Fig. 4.1). The spatial resolution is R2B09, corresponding to an

[1] Polar Stratosphere in a Changing Climate (https://www.polstracc.kit.edu)
[2] Investigation of the life cycle of gravity waves (http://www.pa.op.dlr.de/gwlcycle2/)
[3] Seasonality of Air mass transport and origin in the Lowermost Stratosphere using the HALO Aircraft

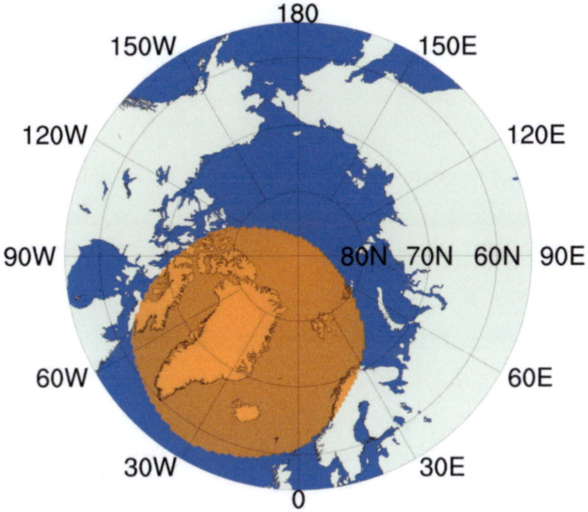

Figure 4.1: Simulation domain for this study.

effective horizontal mesh size of approximately 5 km and 50 vertical levels reaching up to 20 km hence resulting in a mean distance between the model layers of $\sim$ 1000 m in the upper troposphere and $\sim$ 10 m close to the surface. The integration time step is 25 s and all simulations are carried out for a 24 h-period.

For this study, ICON-ART is coupled with the two-moment microphysical scheme (see section 2.4). To obtain realistic rates for the formation of ice crystals, assess the competition between homogeneous and heterogeneous nucleation, and investigate the impact of the artificial seeding aerosol, the nucleation process is parameterized using the formulation after Barahona and Nenes (2009b), as described in detail in section 2.5.

Besides temperature and supersaturation, homogeneous nucleation depends only on number concentration (n_{HOM}) and diameter (d_{HOM}) of small liquid aerosol droplets (Koop et al., 2000). Assuming ubiquitous presence of such particles (Köhler and Seifert, 2015), constant values for $n_{\mathrm{HOM}} = 1000\ \mathrm{cm}^{-3}$ and

$d_{HOM} = 200$ nm are prescribed throughout the atmosphere. To parameterize heterogeneous nucleation, the formulation after Phillips et al. (2013) is applied. In this study, only mineral dust is considered as possible INP, as this species both has effective ice-nucleating properties and prevails over most of the atmosphere. Other species like soot, organic or metallic particles might therefore be neglected in modeling studies (Cziczo et al., 2013).
As even with a relatively high spatial resolution of 5 km, parts of the dynamical structures decisive for cloud formation remain unresolved, a TKE-based Gaussian distribution to account for the sub-grid scale variability of the vertical velocity is applied (Eq. 2.37) after Lohmann et al. (1999), who proposed $f_w = 0.7$ for use in coarse-resolution global models, whereas Rieger et al. (2015) found $f_w = 0.3$ yielding better results in higher resolution ICON-ART simulations of a mid-latitude spring episode compared to measurements. The latter are found to yield reasonable results also for the simulations presented in this study. For consistency, also the formation of cloud droplets is calculated in an aerosol-aware manner by applying the parameterizations described in section 2.5.
For the semi-idealized simulations carried out in this study, instead of considering both natural and seeding aerosol as prognostic species, i. e. describing them with an explicit parameterization for emission and let them undergo atmospheric transport and turbulent diffusion processes, averaged profiles (Fig. 4.2) are prescribed, obtained from a spin-up simulation with ICON-ART covering the entire globe.
Mineral dust over the Arctic originates form sources far off the simulation domain, mainly located in the Sahara. Due to the long-range transport, number concentrations are generally low. In altitudes between 8 km to 13 km, where cirrus formation is expected, henceforth thus termed as *cirrus region*, number concentrations of mineral dust (n_{DUST}) for mode *A* do not exceed 1 l^{-1} to 10 l^{-1}. Such values agree with other studies (e. g. Cziczo et al., 2013). Within the well-mixed, planetary boundary layer (PBL), mineral dust that sedimented from above experiences strong dilution, resulting in low average values of

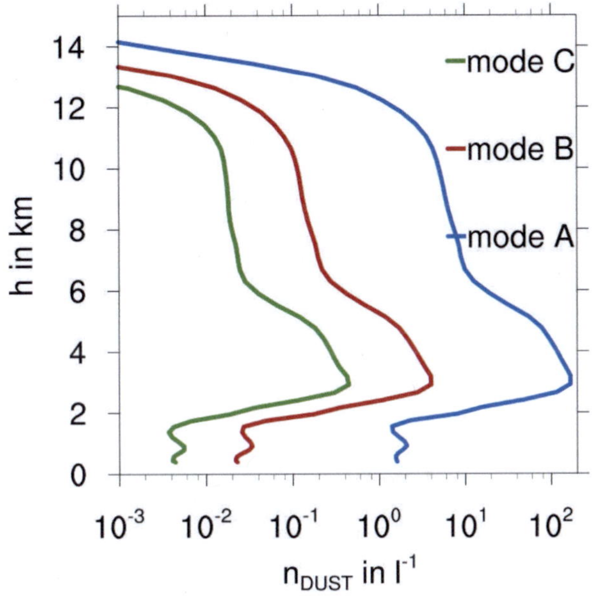

Figure 4.2: Averaged height profiles of number concentrations (n_{DUST}) for the three modes of mineral dust.

n_{DUST} for mode *A* around 1 l^{-1} to 2 l^{-1}. The highest values for n_{DUST} are located just above the PBL between 2 km and 5 km above ground, reaching approximately 120 l^{-1} for mode *A*, and decaying strongly with height beyond. Prior to the date considered for this study, a special synoptic situation led to an unusual, intense warming of the Arctic region north of Europe. Subtropical air was transported poleward due to an intense low-level jet and reached the area north of Svalbard within the outflow of a warm conveyor belt (Binder et al., 2017). Due to low temperatures at the ground of the region forming a layer of cold air, the dust-laden, warmer air masses were forced to rise, causing the maximum mineral dust concentrations at altitudes between 2 km and 5 km.

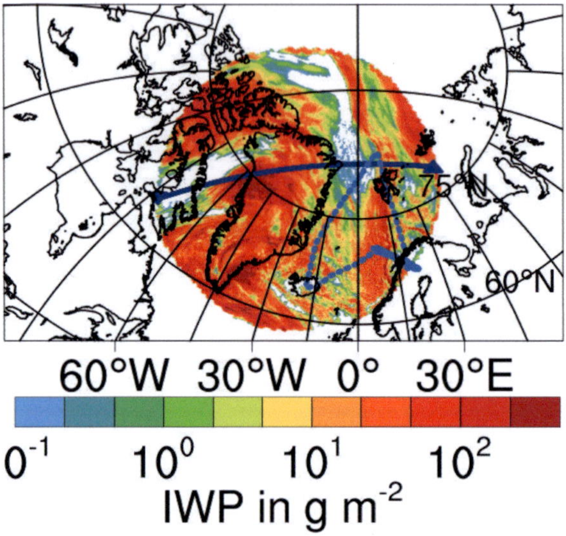

Figure 4.3: Simulated ice water path at 07 UTC. The dark blue line represents the CALIPSO swath crossing the domain from the location with the up- to the downward pointing triangle shown in Fig. 4.4. Light blue lines indicate flight paths (at different times of day) of HALO (dotted) and Falcon (solid).

As mean dust aerosol diameters do not change much with height, a value constant with height for each mode is prescribed. An overview of the initial values for modal parameters of both aerosol species can be found in Tab. 2.1.

4.2 Validation

In contrast to earlier modeling studies investigating the impact of seeding aerosol on cirrus clouds, the present study makes use of an entirely different model setup. Due to the high spatial and temporal resolution, a thorough validation of the simulated atmospheric processes with measurement data is possible. Besides, the present study is one of the few attempts yet of high-resolution modeling of Arctic cirrus clouds. Focusing on the hibernal Arctic, however, makes validation a challenging task. Most satellite sensors are unable to gather data in the absence of solar radiation during polar night, and in-situ measurement campaigns usually are not conducted in winter, presumable by virtue of similar reasons. Providentially, during the PGS campaign, both in-situ as well as remote sensing data was gathered on numerous flights of the two DLR[4] research aircraft HALO[5] and Falcon. Amongst other objectives, the campaign aimed at investigating the structure and composition of the Arctic upper troposphere and lower stratosphere, chemical and physical processes leading to Arctic stratospheric ozone depletion, and transport and mixing processes in the upper troposphere and lower stratosphere (UTLS) at high latitudes (e.g. Khosrawi et al., 2017). In the following, analyses are performed for 20 January 2016, the day on which PGS flight 8 was carried out. In Fig. 4.3, the flight path of HALO is depicted as a dotted blue line.

4.2.1 Extinction coefficient

For a first comparison, data gathered from the CALIPSO[6] satellite (Winker et al., 2007, 2009) is used. The satellite is one component of the A-train satellite constellation (Stephens et al., 2002). With an orbit inclination of 98.2°, daily measurements over the polar regions up to 82° N are available. Aboard the satellite, the CALIOP[7] instrument (Hunt et al., 2009; Powell et al., 2009) is mounted, measuring backscatter at wavelengths of 1.064 μm and 0.532 μm, respectively; thus, as not reliant upon solar radiation, it is able to gain data also during polar night. In the following, the data analyzed is based on the retrieval for the extinction coefficient at 0.532 μm from v3 CALIOP Level 2 Profile Cloud product available through the NASA Langley Atmospheric Science Data Center (ASDC) (`http://eosweb.larc.nasa.gov`) with a horizontal resolution of 5 km.

[4] Deutsches Zentrum für Luft- und Raumfahrt
[5] High Altitude and Long Range Research Aircraft
[6] Cloud-Aerosol Lidar and Infrared Pathfinder Satellite Observations
[7] Cloud-Aerosol Lidar with Orthogonal Polarization

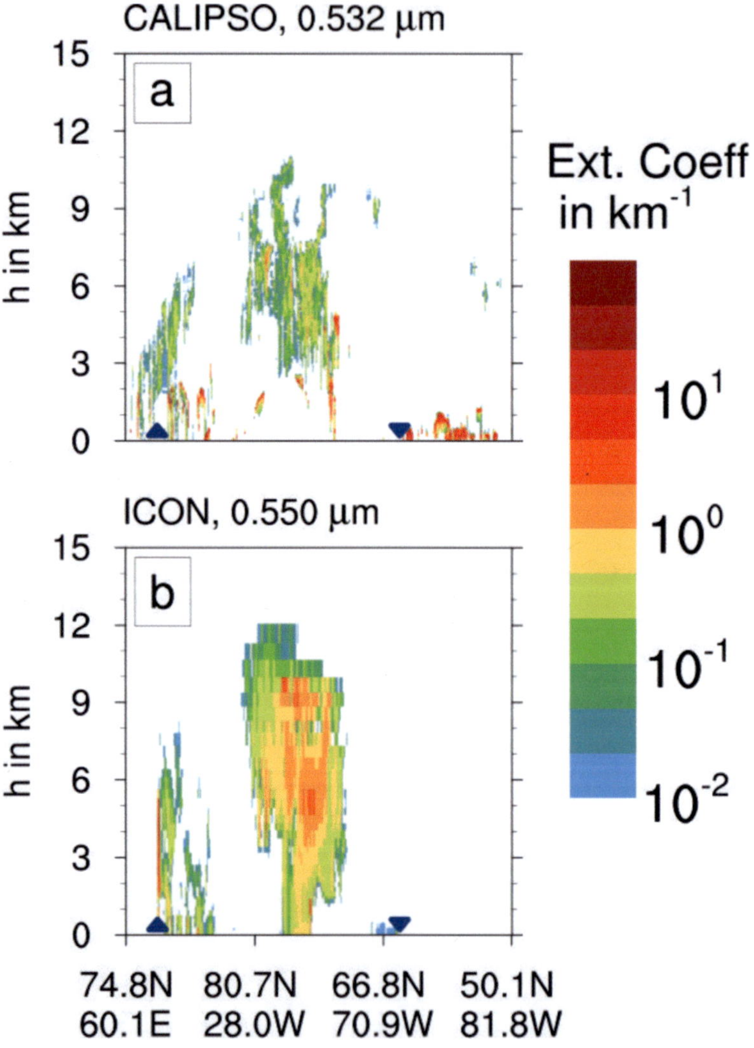

Figure 4.4: Exemplary cross section of extinction coefficient along the dark blue line from up- to downward pointing triangle in Fig. 4.3 of CALIPSO (a) and simulation(b).

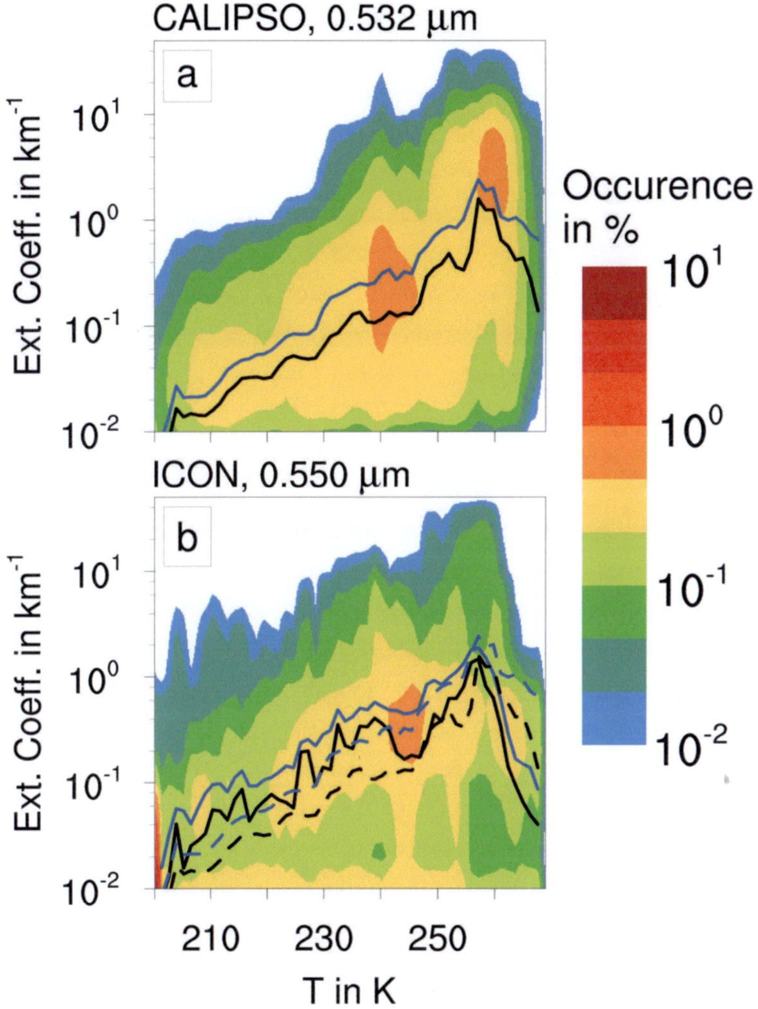

Figure 4.5: Two-dimensional probability distribution of the extinction coefficient with respect to temperature sampled along all tracks of CALIPSO crossing the simulation domain during the entire integration time for CALIPSO (a) and reference simulation (b). Black (blue) lines indicate average (median) extinction coefficient in each temperature bin. In (b), average and median from (a) are shown as dashed lines.

An example of a CALIPSO swath passing the simulation domain is depicted in Fig. 4.3 represented by the dark blue line starting from the upward pointing and ending at the downward pointing triangle. Along this swath, the cross sections of the extinction coefficient are depicted for CALIPSO at 0.532 μm (Fig. 4.4a) and the nearest ICON wavelength 0.550 μm (Fig. 4.4b). In general, both values as well as spatial patterns agree reasonably well between satellite and model. Especially, the top heights of the ice clouds are captured well by the model. At lower altitudes, the model seems to overestimate the extinction coefficient, hinting at either an overestimation of the ice water content (IWC) or rather an underestimation of n_{ICE}. However, it is also possible that uncertainties regarding the retrieval algorithm of CALIOP induce a certain underestimation of the extinction coefficient, as the laser beam is not able to penetrate the cloudy layers above.

A more robust comparison of CALIPSO and the simulation is done by sampling all swaths of CALIPSO during the 24 h integration time of the reference simulation as depicted in Fig. 4.5. Throughout the chosen codomain of both temperature and extinction coefficient, observation and simulation agree reasonably well. However, the model overestimates the frequency of occurrence of extinction coefficient larger than 1 km^{-1} at temperatures below 230 K, i. e. at higher altitudes. Moreover, also very small values about 10^{-2} km^{-1} occur more frequently in the simulation. Deviations between observation and simulation at higher temperatures are attributed to the issue mentioned above. In a statistical mean, the model slightly overestimates the extinction coefficient within the temperature range from 200 K to about 235 K systematically.

4.2.2 Supersaturation

The very basis of ice crystal nucleation is supersaturation with respect to water vapor. To simulate realistic cloud formation, a crucial requirement of atmospheric models is to generate realistic distributions of humidity. However,

validation is challenging. While there are in-situ measurements of humidity in the UTLS region with high time resolution and accuracy available (e.g. Krämer et al., 2009), they are limited to the flight tracks of their carriers. In contrast, data obtained from large-scale satellite observations suffer from a limited vertical resolution (Lamquin et al., 2012).
During PGS, HALO was equipped with the GLORIA[8] instrument. GLORIA is a limb-imaging Fourier transform spectrometer. Details about GLORIA are discussed by Friedl Vallon et al. (2014) and Riese et al. (2014). Two operation modes of GLORIA were applied during PGS: the 'chemistry' mode for high-spectral but medium-spatial sampling and the 'dynamics' mode for medium-spectral, high-spatial sampling. For the comparison in this study, the cloud index (Spang et al., 2004) and retrievals of water vapor (H_2O) and temperature based on 'chemistry' mode measurements are used. This observational mode is particularly useful for measurements of weak trace gas signatures, especially under cloud-affected conditions. The GLORIA 'chemistry mode' measurements during PGS are discussed in detail and validated by Johansson et al. (2018). The specific ICON-ART simulation used for comparison with GLORIA here was initialized for the day of the observations at 00 UTC. Thus, the GLORIA data are compared with the model data after a forecasting period of 15 to 19.5 hours, depending on the particular part of the flight.
The cloud index (Spang et al., 2004) is a spectral color ratio applied to infrared limb observations and is sensitive to increasing opaqueness in the presence clouds. Usually, cloud index values close to 1 indicate opaque conditions due to optically dense clouds, while cloud index values above 4 indicate cloud-free conditions. Values in between indicate partly cloud-affected conditions due to the presence of partly transparent clouds or measurements with the instrument's field-of-view covering clouds only partly. Applied threshold values vary for different sensors. During the flight analyzed here, PSCs[9] were furthermore present above flight altitude, resulting in low cloud index values

[8] Gimballed Limb Observer for Radiance Imaging of the Atmosphere
[9] polar stratospheric clouds

around 2 close to flight altitude, which are not indicative of cirrus clouds. Furthermore, the signal of cloud-free and cloud-affected air masses below flight altitude, which is the focus here, is altered by the superimposed faint PSC signal from above. However, while no strict threshold value can be determined here, the GLORIA cloud index data for PGS08 shows clearly distinguished cloud patters below flight altitude and is therefore well suited for comparisons with ICON-ART with respect to the locations and extents of clouds.
The water vapor volume mixing ratio and temperature are retrieved from the GLORIA observations involving a radiative transfer model and inversion algorithm (Johansson et al., 2018, and references therein). Furthermore, relative humidity is calculated from the GLORIA temperature and water vapor data. With GLORIA measurements affected by opaque clouds, i. e. cloud index close to 1, no atmospheric parameter retrieval is possible. Hence, gaps result in the GLORIA temperature, water vapor and ice supersaturation data in the presence of clouds.
As the GLORIA measurements during this flight were affected by polar stratospheric clouds and cirrus clouds, moderate vertical resolutions of typically 0.5 km to 1 km were achieved for the retrieved profiles. Data points with vertical resolutions lower than 2.0 km are excluded here.
Figure 4.6a shows the GLORIA cloud index data for the selected flight section. Gaps in the data correspond with a flight section where GLORIA was operated in the 'dynamics' mode (about $17:00\ UTC$ to $18:15\ UTC$) and time intervals of calibration measurements. Figure 4.6b shows a cloud mask generated from the model data along the geolocations of the GLORIA data, including also data where gaps are present in the measurement. At the beginning of the cross-section, the GLORIA data show a dense cloud, i. e. cloud index close to 1, with a cloud top decreasing from about 10 km to below 6 km at $15:40\ UTC$. A similar cloud structure is found in the model data, beginning above flight altitude and ceasing at about $16:00\ UTC$. While the shape of the cloud top is similar, the systematic difference of the cloud top altitudes is probably due to a slightly different horizontal location of the cloud system in

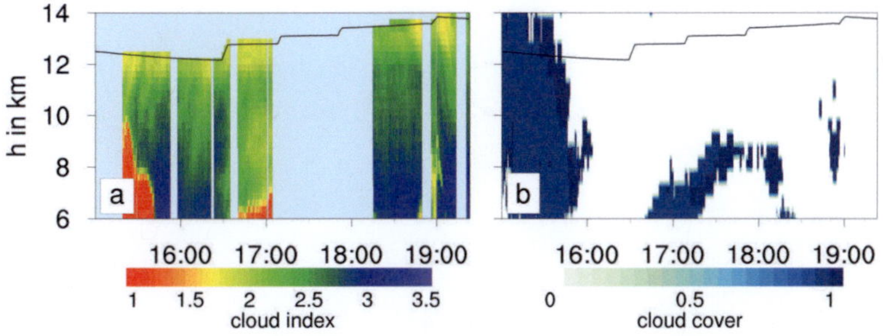

Figure 4.6: (a) Vertical cross-section of cloud index derived from GLORIA measurements during PGS08 on 20 January 2016. Low cloud index values indicate opaque conditions due to cloud particles along the instrument's line of sight. (b) Cloud mask derived from the model simulation at the geolocations of the measurement (0 = no cloud / 1 = cloud present). Black line in both panels: HALO flight altitude. Times in UTC.

the simulation. Note that cloud index values close to 2 along the flight track are the consequence of PSCs at higher altitudes, affecting the GLORIA limb views. Thus, no cirrus cloud signal is expected in the ICON-ART data here. Another dense cloud below about 7.0 km to 7.5 km stretching in the GLORIA data from shortly after $16:30\ UTC$ to beyond $17:00\ UTC$ is reproduced in location and horizontal extent by the model, while the cloud ceiling shows a slightly different pattern. In the model data, the cloud system further arches up to about 9 km in the measurement gap and declines below 6 km at $18:30\ UTC$. Finally, an optically less dense cloud is found in the GLORIA data in a narrow time interval from about $18:45\ UTC$ to $19:00\ UTC$ below 9 km. In the same time interval, the forecast data shows a similar cloud structure at about 1 km to 2 km higher altitudes.

Figure 4.7 shows GLORIA data of water vapor volume mixing ratio and relative humidity (i.e. ice supersaturation) together with the corresponding model data. The black dots indicate the location of the thermal tropopause derived from the GLORIA and ICON-ART temperature data, respectively. It is defined as the lowest altitude at which the temperature lapse rate decreases to $\leq 2\ K$ km^{-1} and remains below this value for a range of at least 2 km

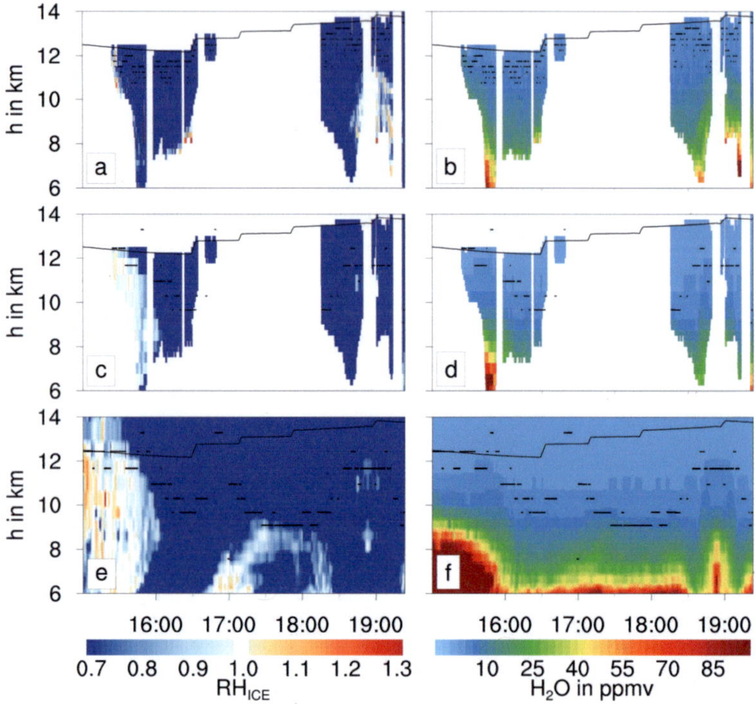

Figure 4.7: Cross sections of relative humidity (i.e. supersaturation versus ice phase) (a) and water vapor volume mixing ratio (b) derived from GLORIA measurements during the HALO flight PGS08 on 20 January 2016. Relative humidity (c) and water vapor volume mixing ratio (d) simulated by the model and filtered for geolocations where GLORIA data are available. Full vertical cross section of the same parameters from the model simulation (e,f). The black lines represent the flight altitude. Black dots indicate the location of the thermal tropopause derived from the GLORIA data and model simulation, respectively. Times in UTC.

(WMO, 1957). The GLORIA and ICON-ART data consistently show typical stratospheric water vapor mixing ratios well below 10 $ppmv$ around and above 10 km (Fig. 4.7b and d/f). While ICON-ART simulates the thermal tropopause at altitudes similar to the GLORIA results before 16 UTC, a systematically lower thermal tropopause by 1 km to 2 km is found in the simulation in the remaining time interval. The simulated increase of water vapor at tropospheric

altitudes below 9 km between 15 : 45 UTC to 16 : 30 UTC agrees very well with the observation. Furthermore, the simulated increase of water vapor towards tropospheric values below 10 km after 18 : 15 UTC shows a similar pattern when compared to the GLORIA data.
The comparison of relative humidity derived from the GLORIA and model data is shown in Fig. 4.7a and c/e. Relative humidity well below 1 (blue) is consistently found in regions characterized by low stratospheric water vapor volume mixing ratios (Fig. 4.7b and d/f) and free of clouds (Fig. 4.6). In the close vicinity of clouds seen in the GLORIA data, relative humidity consistently increases towards 1 and higher in the observation. Similarly, the model simulates enhanced relative humidity in regions covered by clouds and their close vicinity. The slightly different positions of cloud systems in observation and forecast are thereby reflected in the relative humidity distributions (e.g. a region covered by a cloud and characterized by enhanced supersaturation in the model cross section before 16 UTC).
Overall, the ICON-ART simulation shows a high degree of agreement with the GLORIA observations in context of the forecast period. All tropospheric cloud systems found in the observation can be identified clearly in the simulation. Differences in the locations and fine structures of the clouds are explained by the distinct dynamics inherent to the forecast. While the thermal tropopause is mostly located about 1 km to 2 km lower in the simulation, the observed absolute water vapor volume mixing ratios are reproduced well. In context of the slightly different locations and patterns of cloud systems in observation and simulation, relative humidity simulated by the model is also in agreement with the GLORIA data.

4.2.3 Vertical velocity

A crucial atmospheric quantity determining the formation of clouds is vertical velocity, which controls the competition between homogeneous and heterogeneous nucleation (Barahona et al., 2017). Both the magnitude

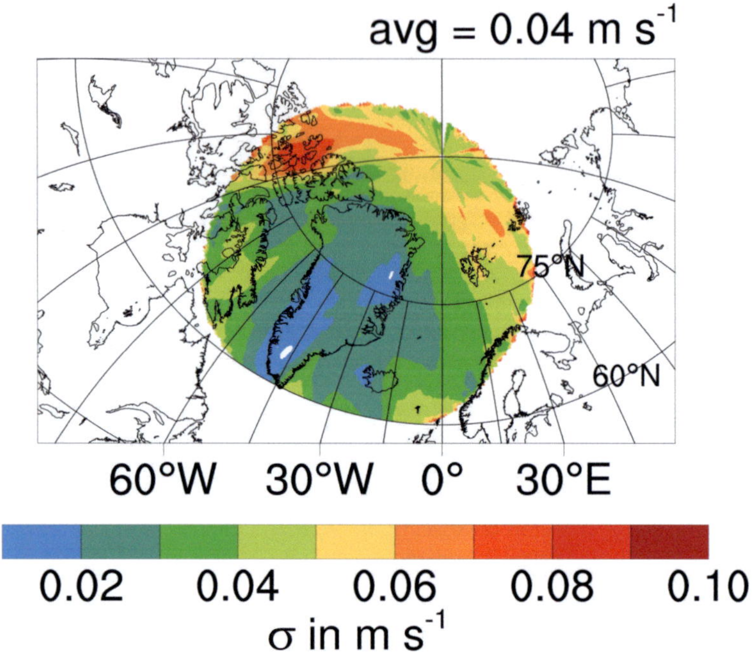

Figure 4.8: Standard deviation of vertical velocity σ_w in m s^{-1} at 250 hPa calculated from a 5 km-ICON-ART simulation averaged over 100 km × 100 km.

and variability of vertical velocity in atmospheric models depends on their horizontal resolution (Rauscher et al., 2016). Fig. 4.8 shows the horizontal distribution of σ_w at 250 hPa for ICON-ART simulations with an effective horizontal resolution of 5 km. The values of σ_w are calculated based on means of areas 100 km × 100 km-wide. On average, σ_w is 0.04 m s^{-1}, over Greenland, values are lower, and north of Canada, values reach up to 0.08 m s^{-1}. These results agree very well with annual means of a 7 km-simulation averaged over 1°-areas by Barahona et al. (2017).

During POLSTRACC, Falcon was equipped with a Doppler wind lidar based on coherent detection with a laser wavelength of 2.02254 μm, capable of

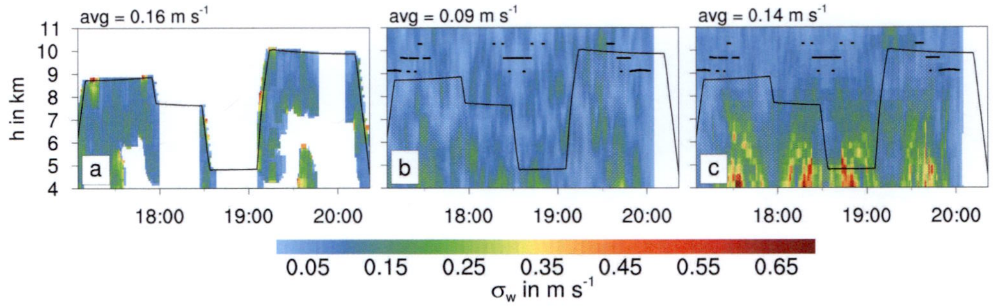

Figure 4.9: Cross section of σ_w for Falcon flight PGS08. a) Measurement of 2μm Wind Lidar; b) ICON, grid scale c) ICON, grid scale with $f_w = 0.3$ in Eq.2.37. The black lines represent the flight height, dotted areas in panels b and c are concurrently covered by the measurements. For details, see text.

measuring vertical velocity in the atmosphere. The standard deviation of measured w along the flight path of Falcon flight PGS08 is depicted in Fig. 4.9a. The black line represents the flight height. In white areas, no measurements were performed, most likely due to clouds being present there preventing further downward penetration of the laser beam. To capture the rather small scale fluctuations of w crucial for cirrus formation, σ_w was calculated with respect to boxes of $\sim$ 1 km $\times$ 1 km $\times$ 1 km. Over large parts, σ_w is small with values around 0.05 m s^{-1}, indicating only minor fluctuations of w. Nevertheless, an increase of σ_w to values larger than 0.30 m s^{-1} is found with decreasing height and especially just above white areas, where clouds may be present, thus representing the upper end of updrafts, e. g. at 18:30 UTC, $h = 5$ km or at 19:45 UTC, $h = 4$ km – 6 km. This seems to be in good agreement with measurements of σ_w in cirrus clouds, although performed at lower latitudes and not in winter (Quante, 2006; Kalesse and Kollias, 2013). Areas with high values of σ_w exist also at altitudes above 8 km. Here, the aircraft may have crossed the tropopause. In Fig. 4.9b, the same calculations for the grid-scale w of the simulation is shown. On average, σ_w is smaller throughout the area. Patterns with peaks of σ_w in Fig. 4.9a are adumbrated, but underestimated by a factor of at least two. Obviously, the model cannot reproduce the small scale fluctuations of w due to the horizontal resolution of approximately 5 km $\times$ 5 km still not being high enough.
To circumvent this issue and still obtaining information on the unresolved, i. e. sub-grid scale, variability of w, its standard deviation σ_w can be parameterized as a function of TKE (Eq. 2.37) (Lohmann et al., 1999), as sub-grid scale variability of w usually is dominated by turbulence.

$$\sigma_w = \sqrt{\sigma_{GRID}^2 + \sigma_{TKE}^2} \tag{4.1}$$

Lohmann et al. (1999) proposed $f_w = 0.7$ for use in coarse-resolution global models, whereas Rieger et al. (2015) found $f_w = 0.3$ yielding better results

in higher resolution ICON-ART simulations of a mid-latitude spring episode compared to measurements.
In Fig. 4.9c, instead of the standard deviation of w based on mean grid scale w, the combination of the grid-scale variability of w and the contribution of the TKE-based formulation of σ_w (Eq. 2.37) with $f_w = 0.3$, are combined according to Eq. 4.1. This choice of f_w captures the areas of enhanced small scale fluctuations of the observed w much better. However, $f_w = 0.3$ still slightly underestimates the observed σ_w at the cloud tops.

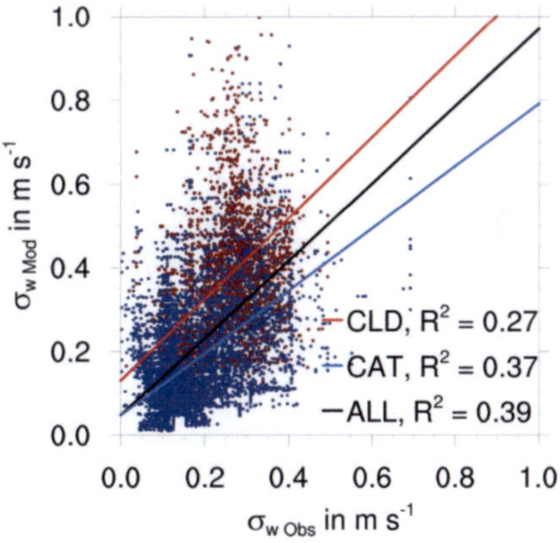

Figure 4.10: Comparison of σ_w for simulated ($f_w = 0.3$ in Eq.2.37) and observed vertical velocity. The lines represent linear regression of all points (black), cloud-free regions (blue), and in-cloud regions (red). For details, see text.

To analyze the impact of using the TKE-based parameterization, a more quantitative comparison is presented in Fig. 4.10. The linear regression of all values is represented by the black line. With a reasonable correlation coefficient of $R^2 = 0.39$, a good correlation is found. However, as GLORIA is unable to penetrate cloudy layers, where high values of σ_w are expected, mostly cloud-free (CAT) regions are observed, depicted in blue in Fig. 4.10. For those points, the model underestimates σ_w, indicating that both the grid-scale fluctuation of w as well as the strength of turbulence characterized by

TKE are too small. In contrast, for the cloudy (CLD) grid points marked in red, the simulated fluctuations are stronger than the observed ones. This deviation can be attributed to larger uncertainties with the calculation of σ_w from the measurements at the margins of the probed regions.

4.2.4 Ice crystal size distribution

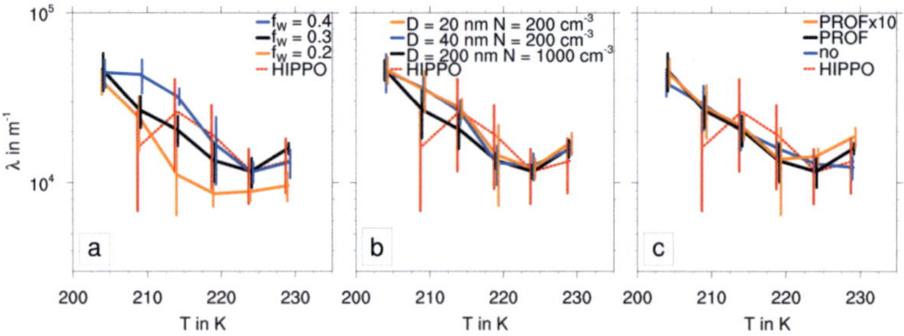

Figure 4.11: Temperature dependence of the slope parameter λ (Eq. 4.2) for HIPPO measurements (red, dotted) and simulations (solid) with different values for f_w (a), sulphate aerosol properties (b), and dust concentrations (c). The vertical bars denote the standard deviation.

From 2009 until 2011, the NSF[10] aircraft campaign HIPPO[11] (Wofsy, 2011) was conducted. During many flight hours (in total over 400 h) of the NSF's Gulfstream V research aircraft, operated by the National Center for Atmospheric Research (NCAR), data was collected over the Pacific region reaching from 67°S to 87° N. In total, 64 flights were carried out, partly also during polar winters. During this campaign, besides other atmospheric quantities, also in-situ measurements of ice crystals were performed. Ice crystal number concentrations were measured with the 2D cloud particle imaging ice probe (2DC) (Korolev et al., 2011). The instrument operates a 64-diode laser array in the size range from 25 μm to 1600 μm at a resolution of 25 μm.

[10] National Science Foundation

[11] HIAPER Pole-to-Pole Observations

The data obtained during Arctic winter (1st October until 31st March) from each of the missions HIPPO 2 - 5 is used here for statistical validation of the simulated ice crystal size distributions. Filtering is performed according to the criteria deployed by Wu et al. (2017). To avoid distortion caused by shattering of ice crystals at the probes' inlets, only ice crystals larger than 75 μm are taken into account. A common method to verify simulated ice crystal size distributions, is fitting both simulated and measured data to a Γ-distribution (Eq. 4.2) and comparing the slope parameters λ (Heymsfield et al., 2008). In Eq. 4.2, D denotes the crystal size, N_0 is the intercept parameter, and μ is the shape parameter that is set to 0. The data is sorted into 25 μm-wide size bins and temperature intervals of 4 K. Cirrus clouds are defined as $n_{\mathrm{ICE}} > 0.1\ \mathrm{l}^{-1}$ at $T < 235$ K.

$$f(D) = N_0 D^{\mu} \exp(-\lambda D) \tag{4.2}$$

The temperature dependence of λ for both measurements and simulation are depicted in Fig 4.11. The overall trend of both observed as well as simulated values for λ, is an increase with decreasing temperature indicating a decrease of the number-weighted ice crystal sizes. Hence, larger values of λ with decreasing temperature indicate higher number concentrations of the ice crystals at lower temperatures, thus revealing their origin as having been formed by homogeneous nucleation. As additionally less water vapor is available at these altitudes for depositional growth compared to warmer, lower layers, ice crystals stay rather small. Performing similar analyses, this correlation was also found in former studies (Heymsfield et al., 2008; Eidhammer et al., 2014; Wu et al., 2017). The model seems to overestimate the mean ice crystal size, or rather underestimate n_{ICE}, as the values of λ for the latter are generally about two to four times smaller compared to the observed values. However, they still lay within the range of the standard deviation from the observed λ. Nevertheless, as the model simulations represent only

a specific case study, deviations between simulations and observations are to be expected, in contrast to e. g. comparing specific measurements along flight tracks for the exactly same time as discussed before (section 4.2.2).
Variation of f_w in Eq. 2.37 has a large impact on the size distribution of the ice crystals in cirrus clouds (Fig 4.11a). As seen before, $f_\mathrm{w} = 0.2$ (orange) and $f_\mathrm{w} = 0.3$ (black), representing only small sub-grid scale fluctuations of the vertical velocity, underestimate the observed values rather strongly, whereas $f_\mathrm{w} = 0.4$ (blue) agrees better with the distributions obtained from the HIPPO data.
For temperatures below 215 K, the observed λ, in contrast to the simulation results, decreases with decreasing temperature, which might be either caused by limited and uncertain measurement data, or rather correspond to very low number concentrations. When the UTLS region is very cold and dry, the cirrus clouds present there contain only few ice crystals (Krämer et al., 2009). Assuming both smaller mean diameters and lower number concentrations of the liquid sulphate aerosol particles only has small impact on the size distribution of ice crystals (Fig 4.11b). Even relatively small concentrations of 200 cm^{-3} still seem to be high enough to not limit the number of ice crystals formed via homogeneous nucleation. Smaller mean diameters of the sulphate aerosol particles result in slightly larger values for λ. Here, both smaller and more ice crystals form, as the smaller ice crystals do not consume as much water vapor and thus do not reduce supersaturation as much as larger ice crystals would do. However, the effect seems do be of rather minor importance.
Fig. 4.11c displays the impact of varied dust concentrations. Besides a 'realistic' profile (black) averaged from the spin-up simulation (Fig 4.2), a case with ten times higher concentrations (orange) and a completely dust-free scenario (blue) are assumed. At temperatures lower than 215 K, hardly any difference occurs, indicating that heterogeneous nucleation is not the dominant formation process of ice crystals, as even ten times higher dust concentrations are still too low to effectively inhibit homogeneous nucleation. In regions with temperatures larger than 225 K, enhanced dust concentrations result in

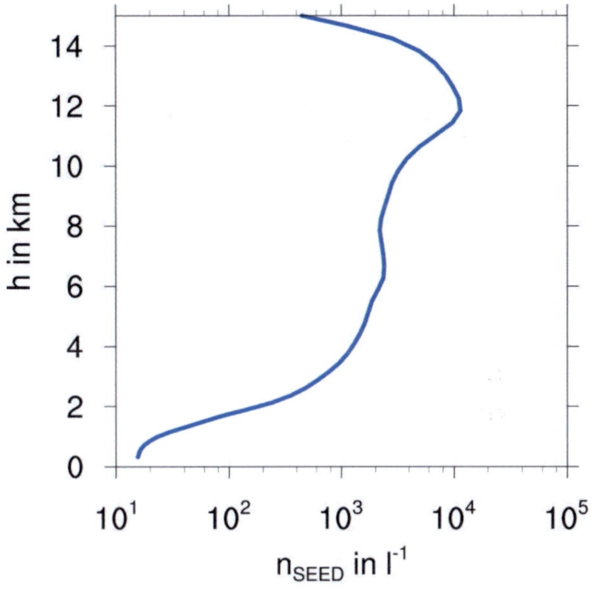

Figure 4.12: Averaged height-profile of number concentrations for the seeding aerosol.

higher number concentrations. Here, homogeneous nucleation occurs more rarely compared to colder regions. Furthermore, dust concentrations are higher than above, resulting in significantly larger ice crystal number concentrations formed from the dust aerosol particles.

In general, the standard deviation of λ is larger for the HIPPO data compared to the simulation. This indicates that due to the horizontal resolution of about 5 km $\times$ 5 km, the model misses part of the small scale variability in cirrus clouds that the 2DC-measurements is still able to resolve.

4.3 Influence of seeding

Various substances have been proposed to be used as INP for cirrus seeding. Ideally, the substance of choice should be very effective at nucleating ice

at temperatures lower than $-20°$ C but ineffective at higher temperatures. Furthermore, it should be nontoxic and affordable. Several studies assumed seeding scenarios with such a 'perfect' substance (e.g. Storelvmo et al., 2014; Gasparini et al., 2017), hence disregarding methods of deployment but rather assuming that the aerosol particles were in place in necessary concentrations. A means of weather modification, i. e. prevention of hail, uses rather expensive silver iodide (Dessens, 1998). Cheaper, but still suitable agents could be Bismuth tri-iodide (BiI_3) (Mitchell and Finnegan, 2009) and mineral dust (Lohmann and Gasparini, 2017). However, not much is known about the ice-nucleation properties of BiI_3 from laboratory studies or even field measurements, whereas mineral dust has been investigated to a much greater extent (e. g. Eastwood et al., 2008; Niemand et al., 2012; Atkinson et al., 2013; Boose et al., 2016). Therefore, the impact of a hypothetical deployment of seeding particles with ice-nucleating properties like mineral dust is examined. As besides the number concentration, also the size of the aerosol particles determines their ice-nucleation properties (Hoose and Möhler, 2012; Gasparini et al., 2017), scenarios with several different mean diameters of the seeding aerosol are assumed.
Commercial aircraft could be a suitable means for delivery, being both affordable and already flying at the desired altitudes. Possible technical solutions could be either carrying the substance dissolved in the jet fuel and burn it to create the seeding aerosol, or by injecting the substance into the engine exhaust forming the seeding aerosol in the jets' contrails (Mitchell and Finnegan, 2009). In this study, this idea of deployment is mimicked. To obtain an average height profile, a spin-up simulation of two weeks was performed, where deployment of the aerosol took place alongside aircraft trajectories at cruise altitude throughout the Northern hemisphere. The trajectories were obtained using traffic way-point information from ADS-B[12] transponders on commercial airliners according to the method described in Gruber et al. (2018).

[12] Automatic Dependent Surveillance - Broadcast

The average height profile of the seeding aerosol number concentrations (n_{SEED}) is depicted in Fig. 4.12. Despite the absence of commercial flights over most parts of the Arctic region, the maximum concentration of seeding aerosol is located at the main cruise altitudes between 10 km and 13 km. These high concentrations result from long-range transport, mostly due to the strong hibernal jet stream. Beneath, the concentration drop about an order of magnitude, remaining largely at a constant value throughout the lower free troposphere. In the PBL, the aerosol is diluted due to turbulent processes and removed from the atmosphere by deposition and washout.
For the scenario referred to as *SEED*, such a prescribed, horizontally homogeneous profile is assumed. In addition, results of a highly idealized and more targeted seeding strategy (*TRGT*) are examined. Here, the same concentrations and aerosol sizes as in *SEED* are assumed to be in place only where homogeneous nucleation would occur, thus avoiding formation of artificial cirrus clouds in supersaturated but cloud-free regions and theoretically achieving the maximum possible elimination of cirrus clouds.

4.3.1 Impact on cirrus clouds

In this section, the impacts of seeding aerosol in the scenarios *SEED* and *TRGT* are discussed. The rate of heterogeneous nucleation occurring in areas with $T < 235$ K averaged over the entire simulation time in the reference simulation (*REF*) is depicted in Fig. 4.13a. In the chosen temperature regime, homogeneous nucleation is possible and hence can be suppressed by heterogeneous nucleation on both natural as well as artificial INP.
In the reference simulation, this occurs at the southern tip of Greenland, north of Canada and in the region of Svalbard, possibly due to topography-induced lee waves leading to fluctuations in w which in turn induce supersaturation.
Fig. 4.13b shows the average ratio of heterogeneous to total nucleation rate, i. e. values close to 1 indicate inhibition of homogeneous nucleation. Comparing to Fig. 4.13a, heterogeneous nucleation is dominant wherever it occurs at

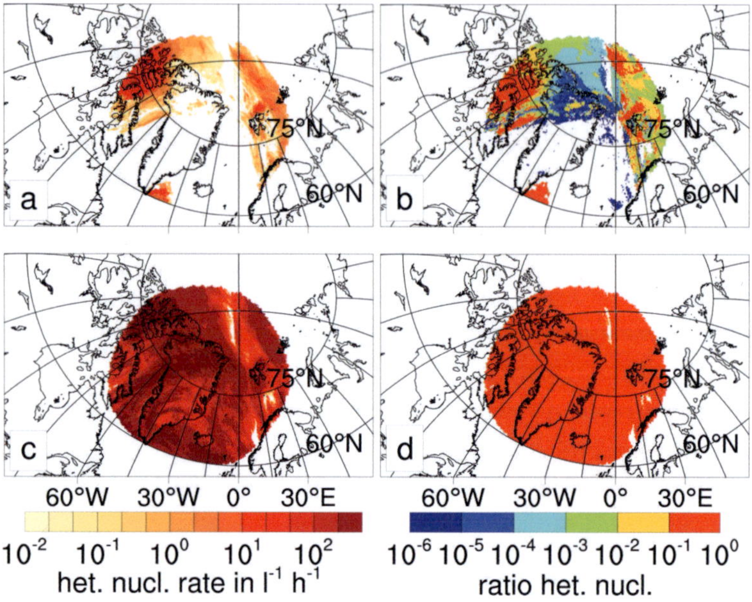

Figure 4.13: Heterogeneous nucleation rate (left) and ratio of heterogeneous to total nucleation rates (right) for reference simulation (top) and *SEED* (bottom) averaged over the entire domain and simulation time.

sufficiently high rates, above approximately 10 l^{-1} h^{-1}. Throughout these regions, seeding cannot have the desired effect, as n_{ICE} will only be enhanced by the artificially introduced INP. Obviously, as heterogeneous nucleation is ubiquitous in the *SEED* scenario (Fig. 4.13c), homogeneous nucleation can be suppressed throughout the domain (Fig. 4.13d); except for small regions north of Scandinavia and close to the North Pole. Here, apparently no ice crystal formation is possible at all.

The primary impact of seeding is the reduction of n_{ICE}. The difference in column integrated n_{ICE} for *SEED* and *REF* in regions with $T < 235$ K averaged over the entire simulation time is depicted in Fig. 4.14a and Fig. 4.14b. Overall, a strong absolute (Fig. 4.14a) and relative (Fig. 4.14b) reduction

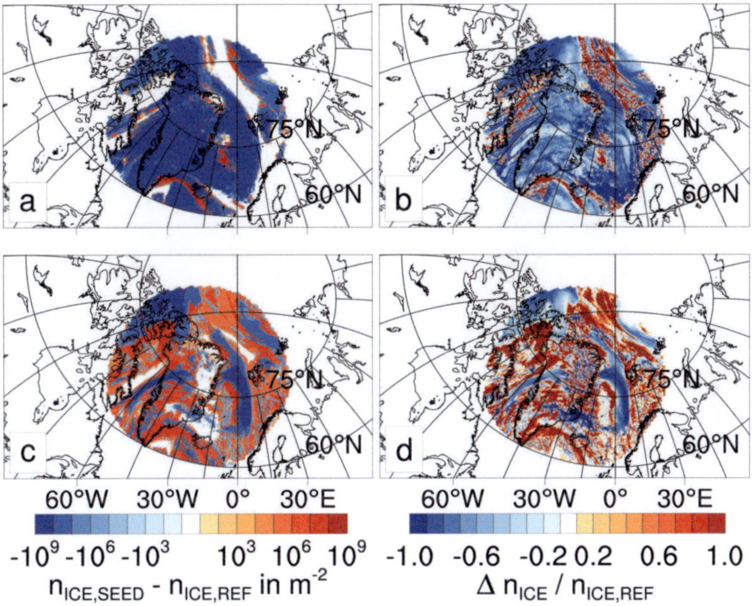

Figure 4.14: Difference in column integrated n_{ICE} for *SEED* and reference simulation in regions with $T < 235$ K (top) and 235 K $< T <$ 273 K (bottom) averaged over the entire domain and simulation time. Left: absolute difference, right: relative difference.

occurs with seeding. However, only minor reductions or even an increase in column integrated n_{ICE} occurs in those regions that were dominated by heterogeneous nucleation already in *REF* (Fig. 4.13b).

A possibly undesirable effect of seeding is formation of artificial cirrus clouds in formerly supersaturated but cloud-free regions (Gasparini et al., 2017). This effect could outweigh most of the reduction of n_{ICE} due to suppression of homogeneous nucleation. In the *SEED* scenario, seeding aerosol is present in vertically varying concentrations but horizontally ubiquitously. Therefore, a strong increase in n_{ICE} can be found in the temperature regime between 235 K and 273 K (Fig. 4.14c and Fig. 4.14d). As here, no homogeneous nucleation can happen and natural INP concentrations are generally low throughout the upper altitudes, injecting artificial INP has a strong impact. However, the column integrated n_{ICE} also decreases on occasion due to enhanced upstream consumption of excess water vapor by the ice crystals formed on the seeding aerosol.

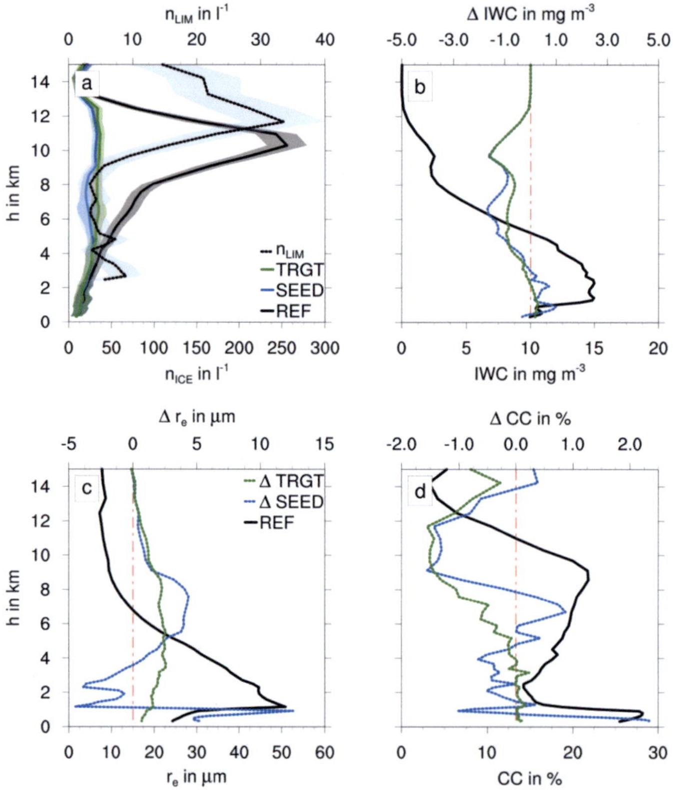

Figure 4.15: Comparison of ice cloud properties averaged over the entire domain and simulation time for reference simulation (black) and seeding strategies *SEED* (blue) and *TRGT* (green). a) n_{ICE}; b) *IWC*; c) r_e; d) cloud cover. Solid lines are absolute values, dotted lines represent differences compared to the reference simulation. Shaded areas indicate standard deviations. Note that in a) only absolute values are depicted, the black dotted line is n_{LIM} for the reference simulation.

Vertical profiles of several cirrus properties averaged over the entire domain and simulation time are depicted in Fig. 4.15a - Fig. 4.15d. In Fig. 4.15a, n_{LIM} is the minimum number concentration of ice crystals formed by heterogeneous nucleation in the reference simulation that still is able to inhibit homogeneous nucleation (Barahona and Nenes, 2009b). At altitudes where high values of n_{ICE} are present, n_{LIM} is lower than about 5 $\mathrm{l^{-1}}$ (at 8 km), increasing strongly with height to up to 35 $\mathrm{l^{-1}}$ at 12 km and decaying further up. This quantity is useful to determine the optimal number concentration of seeding aerosol. For both scenarios, n_{SEED} was chosen in accordance to n_{LIM}, however keeping in mind that the seeding aerosol, being similar to mineral dust, has a freezing efficiency of only about 0.05 (Pruppacher and Klett, 1997). Considering this, a reasonable agreement is found for effective values of n_{SEED} in this study being about 20 times higher compared to former studies assuming 'perfect' INP, i. e. having a freezing efficiency of 1, with optimal values n_{SEED} between 10 $\mathrm{l^{-1}}$ and 100 $\mathrm{l^{-1}}$ (Storelvmo et al., 2013; Gasparini et al., 2017).
In the reference simulation, the highest values of n_{ICE} exceeding 250 $\mathrm{l^{-1}}$ are found at altitudes between 8 km and 13 km (Fig. 4.15a). Both seeding strategies lead to a remarkable reduction of n_{ICE} throughout nearly the entire atmosphere between 3 km and 13 km height. In the levels with the highest values of n_{ICE} in *REF*, seeding leads to ice crystal number concentrations below 50 $\mathrm{l^{-1}}$, corresponding to a reduction to about 20 % of n_{ICE} in *REF*. Interestingly, the reduction of n_{ICE} in *SEED* is stronger than with the *TRGT* scenario that was designed to be of the higher efficiency. Apparently, seeding according to *TRGT* can only inhibit homogeneous nucleation where it is about to occur, whereas with *SEED*, additionally artificial cirrus clouds are formed in regions where supersaturation is too low for homogeneous nucleation, thus on one hand leading to additional cirrus clouds, but concomitantly reducing supersaturation, hence inhibiting homogeneous nucleation further downstream.
In contrast to the strong reduction of n_{ICE}, only minor changes in IWC are found with seeding (Fig. 4.15b). In *REF*, IWC is highest at lower altitudes just above the PBL and strongly decreases with height. In both seeding scenarios,

IWC is reduced in the cirrus region from about 2.5 mg m^{-3} to 1.5 mg m^{-3} corresponding to a reduction of about 60 %. At altitudes between 3.5 km and 9 km, reduction in *IWC* is smaller in *TRGT* compared to *SEED*. Here, as described above, heterogeneous ice nucleation due to seeding and subsequent sedimentation of formed ice crystals leads to decreasing *IWC*.
A slight increase of *IWC* is found in *SEED* at lower altitudes up to 3.5 km, as the seeding aerosol, although only present with low number concentrations, leads to additional artificial ice crystal formation in mixed-phase clouds. However, compared to changes in n_{ICE} at higher altitudes, hardly any change can be found here (Fig. 4.15a) as n_{SEED} is rather low. Nonetheless, within these low altitudes, more humidity is available than above, amplifying growth by deposition as well as glaciation effects, i. e. riming of cloud droplets onto preexisting ice, as well as the Wegener-Bergeron-Findeisen process (e. g. Fan et al., 2011). As in *TRGT*, seeding aerosol is present only where necessary to inhibit homogeneous nucleation, barely any change in *IWC* can be found at lower levels, as heterogeneous nucleation is dominant due to both higher temperatures and higher concentrations of n_{DUST} inhibiting homogeneous nucleation already.
Besides reducing n_{ICE} and *IWC* of ice crystals in cirrus clouds, it is furthermore desirable to increase the ice crystal effective radius (r_{e}) of the remaining ice crystals, as increasing mean ice crystal sizes lead to a decreasing extinction coefficient and thus reduced optical thickness.
In the reference simulation, r_{e} is largest in the PBL, while values in the cirrus region are around 10 μm (Fig. 4.15c). In both seeding scenarios, contrary to former studies, a small increase due to seeding is found. So far, r_{e} was found to either strongly increase (Storelvmo et al., 2013) or rather decrease (Gasparini and Lohmann, 2016). The minor increase of about only 1 μm, compared to up to tenfold values in Storelvmo et al. (2013), can primarily be attributed to the different parameterization used for calculating r_{e} in this study, leading to generally smaller values. Apart from this, although n_{ICE} obviously is reduced strongly (Fig. 4.15a), only a small reduction in *IWC* is found, as values are low

already in the reference simulation (Fig. 4.15b). At lower altitudes between 5 km to 9 km, r_e is larger than above. Here, r_e grows in *SEED*, as the small reduction in *IWC* is accompanied by a much larger decrease in n_{ICE}. As the latter is not happening with *TRGT*, here the increase in r_e is smaller.

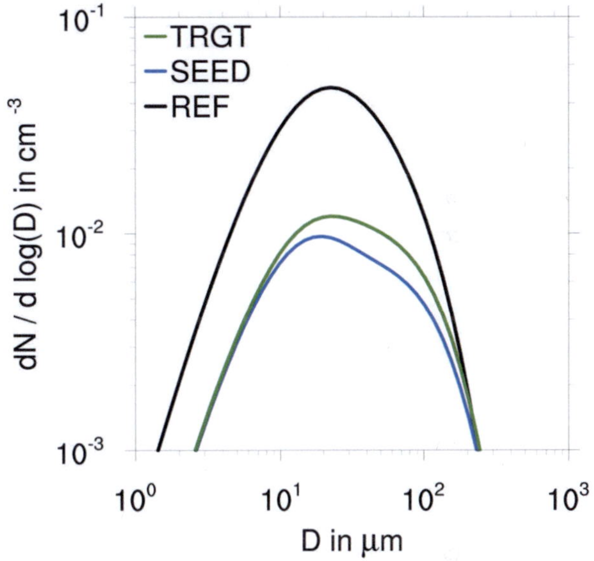

Figure 4.16: Ice crystal size distribution in cirrus levels (8 km - 13 km) averaged over the entire domain and simulation time for reference simulation (black) and different seeding strategies (blue and green).

However, the response of ice crystal properties to seeding is more complex than only the average decrease in n_{ICE} and *IWC* leading to an increase in r_e. Therefore in Fig. 4.16, size distributions of ice crystals, averaged over the cirrus region, are depicted. Consistently with the findings above, the reduction in number concentration is much larger than the average growth in size, as for both seeding strategies, peak number concentrations decrease strongly, whereas the shift to larger mean sizes is hardly evident. Nevertheless, number concentrations of ice crystals with very small sizes decrease strongly with *TRGT* and even more with *SEED*. In contrast, number concentrations of ice crystals larger than about 200 μm are reduced only marginally, and compared to the overall distinct decrease, ice crystals number concentrations with sizes above 20 μm are less strongly reduced, indicating a higher frequency of occurrence of larger crystals, thus broadening the distribution towards larger sizes. This effect is more pronounced when applying the *TRGT* seeding strategy that, by construction, removes more of the small ice crystals arising from homogeneous nucleation.

Former studies pointed at the importance of reducing cirrus cloud cover (*CC*) with seeding (Mitchell and Finnegan, 2009; Storelvmo et al., 2013; Gasparini et al., 2017). However, compared to those, in both *SEED* and *TRGT*, *CC* decreases only marginally in cirrus levels. However, in the simulations presented here, it is nearly impossible to reduce *CC* by seeding, as the integration time of 24 h is too short for sedimentation of smaller ice crystals dehydrating the upper troposphere to a significant extent. Furthermore, a certain contribution from the unseeded lateral boundaries leads to on average diminishing reductions of humidity. In addition, the seeding aerosol number concentrations change with height, but are horizontally homogeneously distributed. Therefore, even when n_{SEED} is high enough to suppress homogeneous nucleation, heterogeneous nucleation still occurs and thus the reduction of cloud cover is small. At altitudes between 5 km and 8 km, cloud cover even increases in *SEED*. As discussed before, at these altitudes heterogeneous nucleation forms artificial cirrus clouds. With regard to achieving maximum cooling, any additional *CC* reduces the primary cooling effect, as outgoing LW radiation is trapped; however, the compensating effect is smaller, the further down the clouds form, as lower-lying clouds emit LW radiation at higher temperatures than colder high-level cirrus clouds.

Injecting artificial INP to seed cirrus clouds is able to efficiently suppress homogeneous nucleation, thus leading to a strong reduction of n_{ICE} for both scenarios considered. However, only small changes are found in *IWC* and r_{e}.

Consistently, also cirrus cloud cover is modified only marginally, contrary to findings of former studies.

4.3.2 Impact on mixed-phase clouds

The response of the liquid phase within mixed-phase clouds is less distinct compared to the one of ice crystals in cirrus clouds. However, in most of the former studies investigating the effects of introducing INP for cirrus seeding, this aspect was not considered (Storelvmo et al., 2013; Storelvmo and Herger, 2014; Gasparini and Lohmann, 2016). In this study, seeding aerosol only acts as INP, so no direct impact on activation of cloud droplets is possible. Nevertheless, due to both the presence of additional INP in lower levels and consequential changes in atmospheric stability and humidity, second order changes within mixed phase clouds are occurring. Changes in n_{CLD}, LWC and r_{e} are small and without a clear trend (not shown). Within the PBL, at altitudes up to about 2 km, a strong decrease in *CC* of up to 4.5 % is found in the *SEED* scenario, corresponding to a reduction by one third (Fig. 4.17). Notably, also the highest altitude where cloud droplets occur at all, decreases about 1 km in the *SEED* scenario.

In the idealized *TRGT* scenario, the effects are smaller and less distinct. Especially, instead of a reduction in *CC* at very low levels, even a small increase is found. This suggests that these secondary effects of seeding are due to introducing additional INP changing ice crystal properties and frequency of occurrence in respective levels.

In a former study, although in the global mean, also a decrease in liquid clouds was found and attributed to increased convective activity due to seeding cirrus clouds (Gasparini et al., 2017). However, during Arctic winter, characterized by low temperatures and lack of incoming solar radiation, convective activity is low (Fig. 4.18a) and the troposphere is rather stably stratified (Fig. 4.18b). In contrast to Gasparini et al. (2017), convective activity in the *SEED* scenario is even slightly reduced and stability is increased within the PBL.

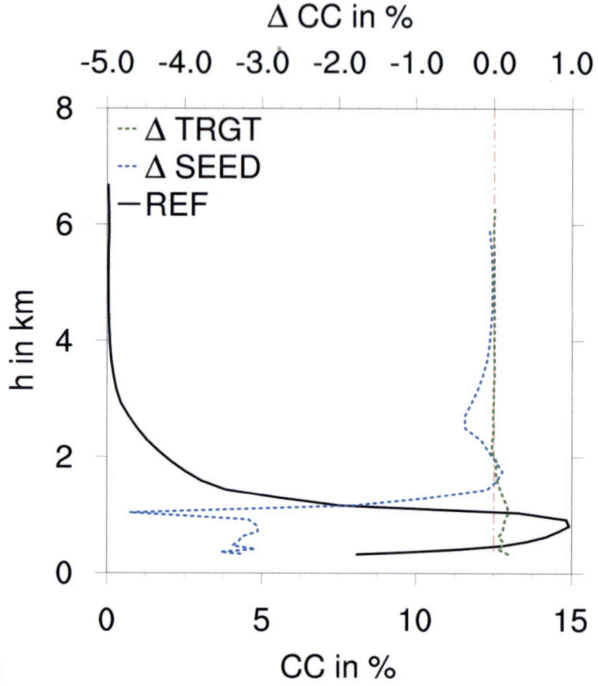

[h!]

Figure 4.17: Comparison of liquid (mixed-phase) cloud cover averaged over the entire domain and simulation time for reference simulation (black) and seeding strategies *SEED* (blue) and *TRGT* (green). The solid lines shows absolute values, dotted lines represent differences compared to the reference simulation.

However, a slight decrease in RH_{ICE} of about 0.01 to 0.02 occurs. Although this effect is small, it hints at enhanced deposition rates onto ice crystals present due to both sedimentation from above and in-situ nucleation, hence reducing supersaturation by consumption of excess water vapor. This in turn reduces the frequency of activation of cloud condensation nuclei.

Furthermore, as already seen for the altitudes above (section. 4.3.1), riming, i. e., freezing of cloud droplets onto ice crystals effectively removes cloud droplets. As this process occurs regardless of the cloud droplet

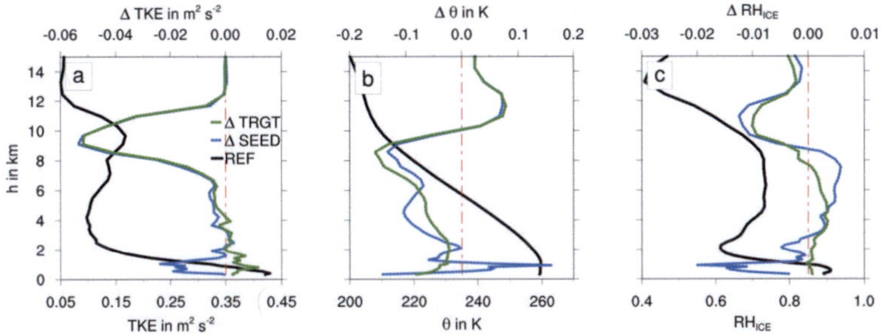

Figure 4.18: Changes of TKE (a), potential temperature (b) and RH_{ICE} (c) averaged over the entire domain and simulation time for reference simulation (black) and seeding strategies *SEED* (blue) and *TRGT* (green). Solid lines are absolute values, dotted lines represent differences compared to the reference simulation.

size, hardly any change in droplet size distributions is found (not shown). Except for a total reduction in number concentration, only small changes are found with the *SEED* scenario. In the idealized *TRGT* scenario, where in absence of homogeneous nucleation no seeding aerosol is present in low altitudes, consequently the size distribution remains merely unchanged. Longer integration time of the simulations and hence mounting redistribution of ice crystals from higher altitudes due to sedimentation is expected to reduce liquid cloud cover even further.

The largest and most distinct response to seeding of the liquid-phase within mixed-phase clouds is a strong reduction of cloud cover within the PBL. This occurs only in the *SEED* scenario and is caused by enhanced riming of cloud droplets onto ice crystals formed due to the presence of seeding aerosol at those altitudes.

4.3.3 Sensitivity to concentration and size of seeding aerosol and background conditions

Both amount and size of the seeding aerosol particles crucially determine the response of both cirrus and mixed-phase clouds on seeding. The efficiency of INP increases strongly with their surface area (Hoose and Möhler, 2012) as the latter increases quadratically with particle size. However, with respect to seeding of cirrus clouds, finding the optimal size of the seeding aerosol particles is difficult. Using larger, more efficient particles, n_{SEED} can be lower than with smaller particles. Hence, as larger particles sediment faster with the fall speed increasing approximately quadratically, too, both increasing efficiency and increasing sedimentation might compensate each other (Gasparini et al., 2017).

However, as in this study, the aerosol particles neither get advected or sediment, nor are scavenged during the nucleation process, n_{SEED} remains constant. This, to some extent, mimics a consecutive injection of seeding aerosol to retain an already built up homogeneously distributed profile.

Table 4.1: Seeding aerosol mean diameters as assumed in Fig. 4.19 and Fig. 4.20.

Simulation	$d_{0,\mathrm{SEED}}$ in μm
DIA1	0.5
DIA2	1.5
DIA3	6.7
DIA4	14.2

Despite the higher efficiency of larger particles, from a technical implementation point of view, smaller particles and thus less total mass might be favored. Therefore, the impact of varying d_{SEED} as listed in Tab. 4.1 is examined. As the choice of values for d_{SEED} is somewhat arbitrary, the same values as for mineral dust are assumed. In addition, the effect of very small ($d_{\mathrm{SEED}} = 0.5\ \mu$m) particles is investigated. For simplicity and better comparison, peak concentrations of $n_{\mathrm{SEED}} = 10^3\ \mathrm{l}^{-1}$ are assumed for all scenarios.

In Fig. 4.19a, averaged vertical profiles of n_{ICE} in response to seeding with changing d_{SEED} are depicted. As expected, reduction in n_{ICE} within the altitudes where cirrus clouds are present, is largest for the larger d_{SEED}. However, the largest aerosol particles with $d_{\mathrm{SEED}} = 14.2\ \mu$m (red) seem to be

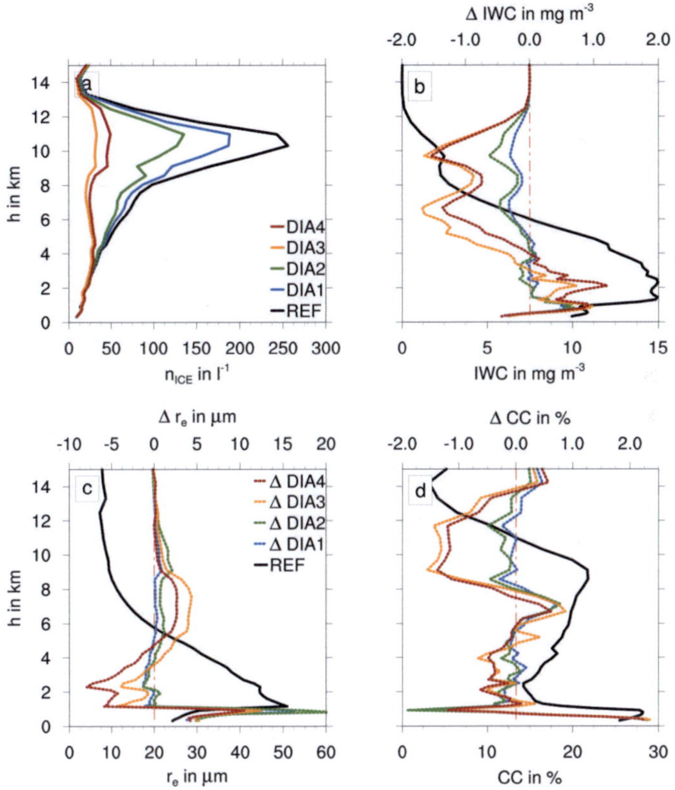

Figure 4.19: Comparison of ice cloud properties averaged over the entire domain and simulation time for reference simulation (black) and simulations with $n_{\mathrm{SEED}} = 10^3$ l^{-1} and various d_{SEED} (colors). a) n_{ICE}; b) *IWC*; c) r_{e}; d) cloud cover. Solid lines are absolute values, dotted lines represent differences compared to the reference simulation. Note that in a) only absolute values are depicted.

slightly less efficient than the second largest ones ($d_{\mathrm{SEED}} = 6.7\ \mu$m, orange) with respect to reducing n_{ICE} (Fig. 4.19a), *IWC* (Fig. 4.19b) and *CC* (Fig. 4.19d) as well as increasing r_{e} (Fig. 4.19c) in the cirrus region. As d_{SEED} is more than doubled from *DIA*3 to *DIA*4, the nucleation effectivity increases by a factor of four. As furthermore n_{SEED} is the same in both scenarios, consequently four times as many ice crystals can be formed in *DIA*4 compared to *DIA*3, leading to a smaller net reduction in n_{ICE}. Apparently, for *DIA*4, the prescribed n_{SEED} is too large, resulting in a beginning over-seeding. As the excess water vapor is distributed on more, and initially larger, ice crystals compared to *DIA*3, their subsequent growth is reduced.

In contrast, with smaller d_{SEED} of 0.5 μm and 1.5 μm, only a moderate reduction of n_{ICE} is found in Fig. 4.19a, as both scenarios are unable to suppress homogeneous nucleation entirely (not shown). Consequently, also the reduction in *IWC* (Fig. 4.19b) and *CC* (Fig. 4.19d), as well as the increase of r_{e} are less distinct compared to seeding with larger particles in Fig. 4.19c.

At lower altitudes, where mixed-phase clouds are present, changes in microphysical properties of both liquid as well as frozen hydrometeors are strongest when injecting the largest seeding aerosol particles. Here, the enhanced water vapor content is not limiting the growth of ice crystals formed by seeding. Consequently, changes within the liquid phase as described in section 4.3.2 are enhanced when using the seeding aerosol particles with the largest d_{SEED}.

In addition, the sensitivity to background conditions was tested in the simulations listed in Tab. D.1 (in appendix D). Varying both diameter and number concentration of sulphate aerosol responsible for homogeneous nucleation of ice crystals within a physical meaningful range seems not to impact the efficacy of seeding (Fig. D.3, Fig. D.4). Also different choices for f_{w} in Eq.2.37 mimicking enhanced or reduced sub-grid scale fluctuations of vertical velocity due to convective activity as input for the nucleation parameterization, have negligible impact on the response of both mixed-phase and cirrus clouds on seeding, although especially n_{ICE} in the reference simulation changes drastically (Fig. D.1, Fig. D.2).

Furthermore, n_{DUST} was changed to either ten times higher concentrations throughout the atmosphere or set to zero assuming complete absence of mineral dust (Fig. D.5, Fig. D.6). As the background dust concentrations are generally low in the UTLS region (Fig. 4.2), even tenfold values are too low for inhibiting

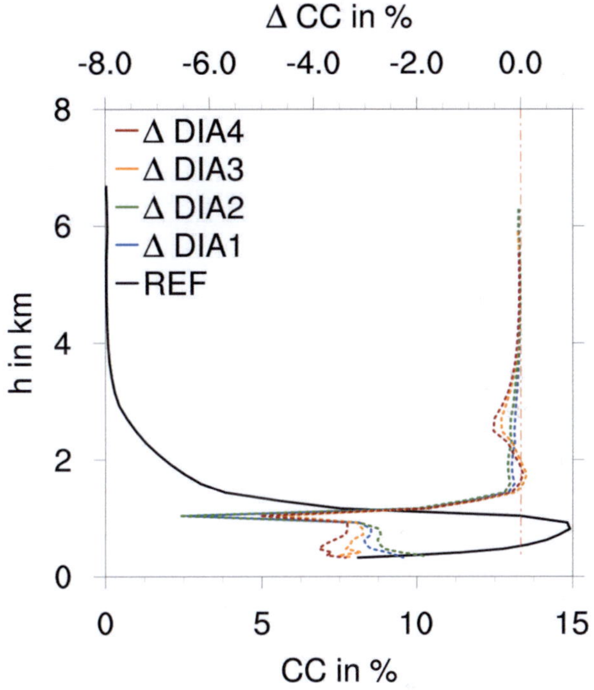

Figure 4.20: Comparison of liquid (mixed-phase) cloud cover averaged over the entire domain and simulation time for reference simulation (black) and simulations with $n_{\mathrm{SEED}} = 10^3\ \mathrm{l}^{-1}$ and various d_{SEED} (colors). The solid lines shows absolute values, dotted lines represent differences compared to the reference simulation.

homogeneous nucleation. Due to the much higher number concentrations of seeding aerosol, in the *SEED* scenario, the latter still is dominant. In the highly unrealistic case of assuming dust-free conditions, response to seeding is much larger for mixed-phase clouds in lower levels. For the ice phase, *IWC* and r_{e} are reduced strongly, whereas *CC* is increased; consequently, the liquid phase is reduced due to consumption of water vapor and riming onto the ice crystals forming on the seeding aerosol particles.

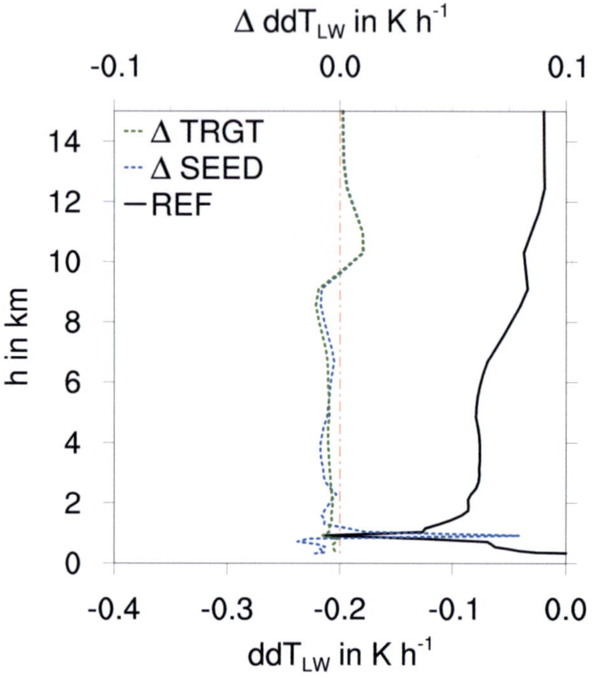

Figure 4.21: Comparison of the LW heating rates averaged over the entire domain and simulation time for reference simulation (black) and seeding strategies *SEED* (blue) and *TRGT* (green). The solid line are absolute values, dotted lines represent differences compared to the reference simulation.

The larger the seeding aerosol particles, the more effective they suppress homogeneous nucleation. However, with too large aerosol particles, artificial cirrus formation becomes more dominant, leading to a smaller net reduction in n_{ICE}. Furthermore, the efficacy of seeding cirrus clouds is relatively independent of the background conditions like sulphate aerosol or mineral dust concentrations, or sub-grid scale fluctuations of the vertical velocity.

4.3.4 Impact on heating rates

Changes in cloud cover, effective radius and *IWC* and *LWC* have an impact on the radiative fluxes that can be characterized by their divergence resulting in changing heating rates. The vertical profile of LW heating rates (ddT_{LW}) averaged over the entire domain and simulation period are depicted in Fig. 4.21. For the chosen day in *REF*, negative LW heating rates are present on average throughout the atmosphere with a maximum cooling in the PBL. Above, the cooling is reduced, increases slightly in the cirrus region at about 9 km. It then decays further up before remaining at a constant value at altitudes above 13 km. Both peaks are just above the major layers of both ice and liquid clouds characterized by cloud cover (Fig.4.15d and Fig.4.17) as well as *IWC* (Fig.4.15b) and *LWC* (not shown). This indicates LW cooling above the cloud top and warming below, characterized by decreasing absolute values of the LW heating rates beneath. Notably, the peak of cooling within the PBL is found with similar values also during Arctic spring and fall (Harrington et al., 1999). The impact of cirrus clouds is in agreement with theoretical studies assuming similar *IWC* of cirrus clouds (Schmidt and Garrett, 2013), but underestimated compared to other studies either using different radiative transfer algorithms (Baer et al., 1996) or rather observation-based calculations focusing on convective cirrus in the tropics (McFarlane et al., 2007). Discrepancies compared to the latter, however, might be explained by much lower temperatures during the Arctic winter of both the surface and the air above, causing only small total LW fluxes.

For *SEED* and *TRGT*, ddT_{LW} is depicted as difference to the values of *REF* in Fig.4.21. For both scenarios, an increase in LW heating is found within the cirrus region, again coinciding well with the maximum of ice clouds in *REF* that are thinned out in both *SEED* and *TRGT*, resulting in a net warming effect. Below, at heights between 2 km and 9 km, stronger LW cooling is found for both scenarios due to the lacking insulating cirrus layer. In the PBL, the strongest response to seeding is found for the *SEED* scenario, where a slight

increase in ice cloud cover (Fig.4.15d) is accompanied by a much stronger reduction in liquid cloud cover (Fig.4.17), resulting in a net removal of clouds and thus causing a rather strong increase in LW heating. Below, just above the ground, cooling as strong as below the thinned cirrus clouds occurs.
In the highly idealized *TRGT* scenario, hardly any changes of cloud properties and occurrence is found within the PBL. The patterns of heating and cooling hence remain merely unchanged here, except for the cooling impact due to the thinning of cirrus clouds proceeding further down.
The reduction in cirrus cloud thickness due to seeding leads to a net LW warming effect within the cirrus regime, whereas moderate cooling is found beneath. Concomitant with the removal of liquid cloud droplets and thus reduction in liquid cloud cover in the PBL, is also a net LW warming effect.

4.3.5 Impact on radiation

The desired effect of injecting aerosol particles into the UTLS region for cirrus thinning is enhancing outgoing LW radiation into space and thus cooling the atmosphere and the Earth's surface beneath. Induced by the short integration time and the limited area, no meaningful estimates on global and annual mean radiative forcing of seeding can be given.
Nevertheless, changes in the radiative fluxes can be examined. The effectiveness of seeding is not only determined by the size of the aerosol particles, but depends also crucially on its number concentrations determining the number of nucleated ice crystals that can deplete excess water vapor and thus, by reducing supersaturation, reduce or even suppress homogeneous nucleation.
Therefore, a number of simulations was carried out varying both n_{SEED} and d_{SEED}. The resulting changes in OLR averaged over the entire integration time and simulation domain are shown as solid lines in Fig. 4.22. The *SEED* scenario discussed in section 4.3.1 and section 4.3.2 with $n_{\mathrm{SEED}} = 10^3\ \mathrm{l}^{-1}$ and $d_{\mathrm{SEED}} = 6.7\ \mu\mathrm{m}$ is amongst the most effective choices, increasing OLR by

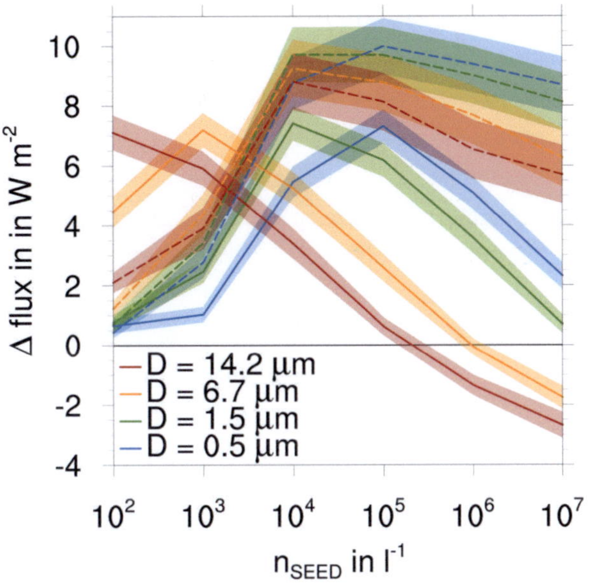

Figure 4.22: Changes in LW net flux at TOA (solid) and SFC (dashed) due to seeding for different combinations of n_{SEED} and d_{SEED} averaged over the entire domain and simulation time. Shaded areas represent standard errors.

7.00 ± 0.31 W m^{-2}. Due to its smaller efficacy, the increase in OLR in the idealized *TRGT* scenario is only 5.80 ± 0.30 W m^{-2}. A major concern in former studies is so-called over-seeding, i. e. a positive radiative forcing resulting from prolonged lifetime or decreasing r_{e} of cirrus clouds due to seeding (Storelvmo et al., 2013; Storelvmo and Herger, 2014; Gasparini and Lohmann, 2016). However, in this study, negative changes in the net OLR at TOA occur only when injecting both larger and very high number concentrations of seeding aerosol particles. Notably, this effect is found only with unrealistic values of n_{SEED} as high as 10^5 l^{-1}, comparable to the number concentrations of ubiquitous sulphate droplets. Most likely, as such n_{SEED} concentrations would be orders of magnitude higher than natural INP concentrations, also

other effects like the direct aerosol effect of the seeding particles might further play a significant role. As expected, when using larger aerosol particles, lower number concentrations are sufficient to achieve a similar enhancement of OLR compared to seeding with smaller particles.
Increasing upward fluxes of LW radiation at the Earth's surface cools the latter. Keeping in mind that the overall cooling effect of cirrus thinning on the climate system can be amplified by exploiting the self-enhancing polar sea-ice-albedo feedback (Holland and Bitz, 2003), this is a desired secondary effect of cirrus seeding. As discussed before, besides only reducing IWC and increasing r_e of cirrus clouds, also mixed-phase clouds are modified, i. e. reduced due to injecting seeding aerosol. Therefore, also the net flux of LW radiation at the Earth's surface (SFC) is examined, depicted as dashed lines in Fig. 4.22. Here, the change in the surface LW net flux in the *SEED* scenario is 4.23 ± 0.39 W m^{-2}. As for *TRGT*, in which hardly any changes in mixed-phase clouds occur, the increase in LW net flux at the surface is only 1.10 ± 0.05 W m^{-2}. For n_SEED up to 10^3 l^{-1}, larger aerosol particle sizes lead to stronger increase in net LW radiation flux at the surface, as liquid cloud cover is reduced more strongly than with smaller sizes. In contrast, for even larger values of n_SEED, smaller seeding aerosol particles are more effective, as predominantly due to enhanced redistribution of water vapor to the ice phase and riming, the reduction in liquid cloud cover is enhanced. For larger, hence more effective, INP, this effect is outweighed by an over-seeding effect, i. e. the reduction in ice crystal r_e resulting from strong increase of n_ICE.
The averaged changes in radiative fluxes for the simulation with modified background conditions mentioned in section 4.3.3 are listed in Tab. D.1. Here, changes in both OLR at TOA and SFC are within the range of the basic *SEED* scenario except for the scenario assuming $n_\mathrm{DUST} = 0$ (*SEED HET2*), as here, due to the absence of natural INP at lower altitudes, the seeding aerosol leads to a great increase in ice cloud cover to values comparable to *REF*, and thus overcompensates the reduction in liquid phase cloud cover. However, as temperatures at low altitudes are too high for homogeneous nucleation,

without the presence of INP, a simple constant mean size of the ice crystals is assumed to diagnose n_{ICE} from the ice mass concentrations of the updated lateral boundaries. Consequently, the choice of this constant mean size strongly determines the results of such simulations.

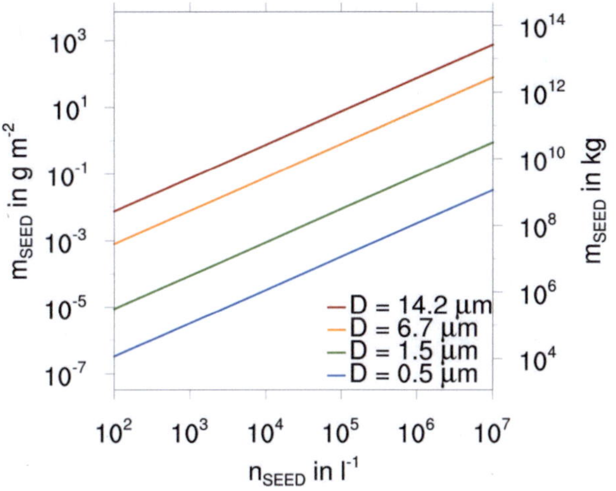

Figure 4.23: Mass of seeding aerosol for different combinations of n_{SEED} and d_{SEED}. Left axis represents column-integrated mass per m^2, right axis represents total aerosol mass for the entire Arctic region (i, e. north of 60°N).

Finally, a rough approximation of the seeding aerosol mass necessary to establish a vertical profile, as depicted in Fig. 4.12, for different choices of n_{SEED} and d_{SEED} is presented in Fig. 4.23. The lines read as either column-integrated mass per m^2 for the left axis, or as total mass necessary to seed the entire Arctic region (i. e. north of of 60° N). In Fig. 4.22, a maximum cooling at TOA of about 7.5 W m^{-2} occurs for several different pairs of n_{SEED} and d_{SEED} relative independently of the actual size and concentration of the seeding aerosol, i. e. with 10^3 l^{-1} and 6.7 μm; 10^4 l^{-1} and 1.5 μm; and 10^5 l^{-1} and 0.5 μm. Those values result in a similar column-integrated aerosol loading of about 10^{-3} g m^{-2} corresponding to about 10^7 to 10^8 kg of seeding material

required for the entire Arctic region. A similar pattern can be found for the threshold towards 'over-seeding', i. e. an increase in OLR at TOA instead of a reduction. Here, the critical column-integrated mass is about 10 g m^2 or rather 10^{11} to 10^{12} kg mass in total. These connections might provide a possible, very simple estimate of the efficacy of seeding Arctic cirrus clouds. However, a less distinct correlation can be found for the maximum change of the surface LW net flux that rather seems to depend on n_{SEED} than on the total mass.
For a wide range of seeding aerosol sizes and concentrations present, the desired effect of increasing both OLR at TOA as well as LW net flux at the surface is found. Additionally, over-seeding occurs only with very high concentrations of the seeding aerosol. Furthermore, rough estimates hint at a direct relation between column-integrated mass of seeding aerosol and their efficacy of cooling.

4.4 Summary

In this study, cloud-resolving simulations were performed with the atmospheric modeling system ICON-ART coupled with a comprehensive and detailed microphysical scheme.
Comparison to measurement data obtained from several sources yield reasonable agreements. Especially the distribution of humidity strongly determining cloud-formation in the upper troposphere, is reproduced well compared to data gathered during the POLSTRACC campaign by GLORIA.
There is a certain lack of measurements of ice crystal properties like size, shape and number concentrations in arctic cirrus clouds, especially during winter. Therefore, ice crystal size distribution obtained from sampling both HIPPO and simulation data were used to adjust natural aerosol concentrations and the probability distribution of sub-grid scale vertical velocity to improve the representation of ice crystal size distribution. The model is applied to clarify the microphysical processes accompanied by injecting artificial aerosol particles into the UTLS region with the aim of modifying cirrus clouds in the

framework of climate engineering. Focusing on the hibernal Arctic region has turned out to be most promising, as here, due to lack of incoming solar radiation, cirrus clouds have only warming effects, and removing or thinning them yield the maximum possible cooling effect. Former studies of cirrus thinning were performed with either a simple proxy for actual seeding, and / or used GCMs with much coarser horizontal resolutions (Storelvmo and Herger, 2014; Muri et al., 2014; Gasparini et al., 2017) focusing on anomalies in global radiative forcing.

In addition, in this study, a somewhat more realistic scenario of injecting seeding aerosol with ice-nucleation properties like mineral dust is investigated by applying a vertical profile of seeding aerosol number concentrations obtained from a separate simulation assuming injection of the seeding material along commercial flight trajectories and subsequent dispersion into the Arctic region.

The primary effect of seeding, i. e. the reduction of ice crystal number concentrations is found to be predominant compared to only small decreases in ice water content and increases of the ice crystal effective radii. Furthermore, minor reductions in cirrus cloud cover are found. Those findings are partly contradictory to former studies that either found large changes in IWC and r_e (e. g. Storelvmo and Herger, 2014), or a slight increase of the latter (Gasparini and Lohmann, 2016).

The impact of cirrus seeding on the longwave radiative fluxes are twofold. Firstly, an increase in outgoing longwave radiation at the top of the atmosphere occurs due to reduced ice crystal number concentrations within cirrus clouds. This effect is present over a wide range of assumed number concentrations and sizes of the seeding aerosol and maximized for number concentrations of 10^3 l^{-1} and rather large particles.

Furthermore, an enhancing secondary effect is found. Mostly due to the enhanced heterogeneous nucleation on seeding aerosol particles present also at lower altitudes, a redistribution of water vapor in favor of the latter and enhanced riming of cloud droplets, originally mixed-phase clouds are reduced

or glaciated. Reduced liquid cloud cover leads to enhanced upward longwave fluxes at Earth's surface, hence generating additional cooling there.
Furthermore, throughout the simulations, the efficacy of seeding on both cirrus and mixed-phase clouds is not sensitive to changes in both natural aerosol concentrations and turbulent activity, except for assuming unrealistic absence of dust.
For low number concentrations of the seeding aerosol, efficacy drops quickly. In contrast, over-seeding is hardly found, with the exception of very large aerosol particles and very high number concentrations close to the concentration of sulphate aerosol droplets responsible for homogeneous nucleation.
Nonetheless, there are major uncertainties related to the process of ice crystal nucleation and a lack of information on the spatial and temporal distribution, the size, and chemical composition of natural aerosol within the hibernal Arctic UTLS region. Even if one neglects the contradictory results of former studies and puts more trust in the presented simulations, showing that seeding cirrus clouds could, in principal, have a certain cooling effect, it is strongly refrained from recommending this method as a means to limit climate change. In particular the readers are discourage to consider this method as a substitute for mitigation.

5 Towards estimating a radiative forcing by thinning of Arctic winter cirrus clouds with ICON-ART simulations

In the previous section 4, a high-resolved, but semi-idealized process study for climate engineering by thinning Arctic winter cirrus clouds was presented. However, several aspect arose that need to be improved. Firstly, due to the rather short integration time of 24 h, the impact of sedimenting seeded ice crystals from the cirrus regions further downward could not be examined in its full strength. Secondly, as the simulations were performed on a limited area, the update of lateral boundary conditions with the same, unseeded data for all scenarios may have an influence on the results that could not be quantified. Especially draining of the upper troposphere due to enhanced heterogeneous ice nucleation rates and subsequent sedimentation is most likely compensated due to the updating procedure.

In order to gain a more comprehensive view on the effects accompanied with injecting aerosol to thin cirrus clouds, and not relying on a single case study, simulations for the entire Arctic Winter 2015 / 2016 are carried out. For this study, as in the previous section, the online-coupled modeling system ICON-ART is used.

Although investigating the impact of cirrus seeding during an entire Arctic winter undoubtedly is far away from being a climate study, the comprehensive and well validated model setup serves to address several important uncertainties of this climate engineering approach. For example,

Storelvmo and Herger (2014) emphasize the necessity of describing seeding aerosol as prognostic species with a full life cycle, i. e. considering both emission as well as removal processes like dry and wet deposition. In doing so, also the distribution and transport of the seeding aerosol in the atmosphere can be addressed.

5.1 Methods

In this study, the Arctic winter 2015 / 2016 is defined as the period from 01 October 2015 until 31 March 2016. However, due to the chaotic nature of the atmosphere, comparing two simulations, each lasting for half a year, is challenging. Introducing a source of disturbance, here in the form of seeding aerosol, will force such a simulation drifting away from a reference. Comparing both simulation states after a certain time, it is nearly impossible to distinguish between those differences caused by seeding aerosol influencing cloud formation and the ones arising from the changed forcing, hence a sort of chaotic noise.

To circumvent this, two options can be considered. A common method applied in many climate studies is performing ensembles, i. e. not relying on a single simulation but rather on a number of simulations, each with slightly modified starting conditions. Like this, statistical noise can be reduced, results are to be interpreted as ensemble means and corresponding standard deviations or spread between the ensemble members.

Another option are experiments with so-called time slices. Here, the temporal period of interest is split into several shorter episodes. Depending on the model itself and the chosen numerical setup, the impact of chaos is negligible when comparing two simulations of the same episode.

Using the latter approach, the Arctic winter 2015 / 2016 is approximated by in total 10 simulations with roughly equally distributed starting dates according to Tab. 5.1, each comprising 72 h. Averages over all of them are considered as mean with respect to the entire period. The starting days for 2016 are shifted

Table 5.1: Starting days for the simulations. In the third column, HALO flights in the PGS campaign are listed.

number	date	PGS
01	01 October	
02	21 October	
03	10 November	
04	01 December	
05	21 December	
06	20 January	PGS08
07	25 January	PGS10
08	26 February	PGS14
09	09 March	PGS17
10	13 March	PGS19

away from a true equal distribution in order to fit better with the dates of the most suitable PGS flights. In the following, single simulations are abbreviated by the scenario considered (*REF* or *SEED*) followed by their number (*1* to *10*). Instead of limiting the simulation domain and relying on prescribing and updating boundary conditions, global simulations are carried out, hence allowing for a fully prognostic treatment of both natural as well as seeding aerosol and not approximate them by prescribed, averaged, and time-independent vertical profiles.

For being able to conduct these simulations with both a much larger domain and a longer integration time, the horizontal resolution as to be decreased to R2B06, i. e. approximately 40 km. A more coarse grid spacing is accompanied with a larger time step for integration. However, as the two moment microphysical scheme used in this study yields unphysical results especially concerning the sedimentation of large hydrometeors and the subsequent lack of interaction of the latter with other hydrometeors, the microphysics are treated

in quasi-implicit manner, as described in section 2.4. In principal, instead of calculating each process rate one after another on a three dimensional array using the updated values of the prior in the next one, the procedure is included within the sedimentation algorithm limited to one height level, hence calculating the sedimentation flux into the current layer, updating all process rates only on this layer and proceeding to the layer beneath. Except for several options depending on the resolution, i. e. parameterized convection and a larger integration time step of 100 s, the model setup used for this study is identical to the one chosen for high-resolution process study (section 4).
In order to start with physically meaningful and realistic conditions, each simulation is initialized with a realistic aerosol loading for mineral dust, sea salt and seeding aerosol. The aerosol distributions are obtained from spin-up simulations with ICON-ART, each starting 10 days ahead of the corresponding dates listed in Tab. 5.1. These simulations are carried out with the same grid spacing, but with a single moment bulk microphysical scheme neglecting the impact of aerosol on cloud formation.
However, as sulphate aerosol decisive for homogeneous nucleation is found to be ubiquitous throughout the atmosphere (Köhler and Seifert, 2015), concentrations and sizes are prescribed being constant in time and space.

5.2 Natural and seeding aerosol

As an example, the distribution of n_{DUST} for the smallest mineral dust mode at 21 October 2015 is depicted in Fig. 5.1. This mode is governing the heterogeneous nucleation of ice crystals due to its high number concentrations. Although hardly any sources for mineral dust are located within the northern Polar circle, here, n_{DUST} partly exceeds 100 l^{-1} in the upper troposphere (Fig. 5.1a). Also at lower altitudes, values are larger than zero are present (Fig. 5.1b), although much smaller than above. In contrast, within the PBL, hardly any mineral dust is found over the Arctic (Fig. 5.1c).

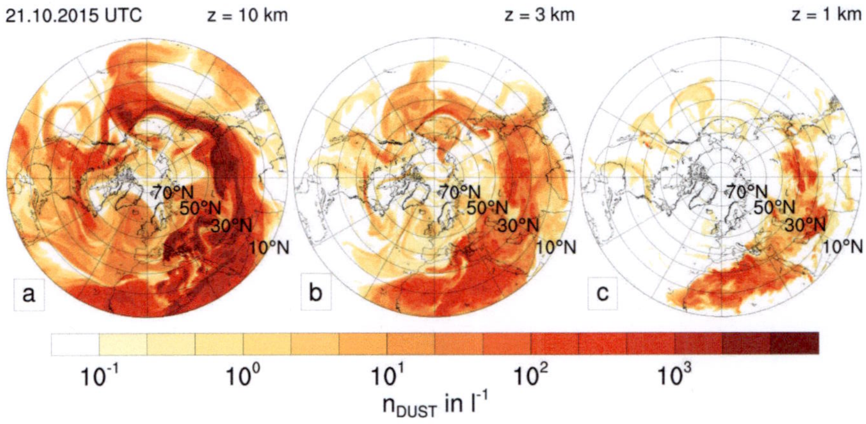

Figure 5.1: Exemplary number concentration of mineral dust (mode A) after 10 days of spin-up for 21 October 2015 in altitudes of 10 km (a), 3 km (b), and 1 km (c).

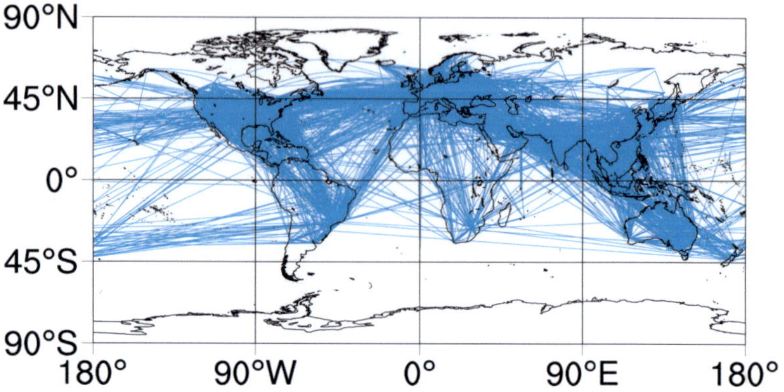

Figure 5.2: Globally distribution of flight trajectories sampled over 24 h used for inject the seeding aerosol into the atmosphere.

The main source regions for mineral dust are found within the Sahara, Northern America and in Central and East Asia. Here, n_{DUST} is largest at low altitudes (Fig. 5.1c). From there, convection and atmospheric currents transport the aerosol particles partly also to higher latitudes. Impressively, planetary wave structures within the jet stream can be identified in form of enhanced concentrations of mineral dust between 20°N and 60°N surrounding the globe (Fig. 5.1a).

These patterns are similar for the other situations considered (not shown); except for a period lasting from end of December 2015 until end of January 2016, where a strong warming event occurred in the Arctic North Atlantic (Binder et al., 2017), already discussed in section 4.1.

Hypothetically, various means of injecting the seeding aerosol into the desired altitudes and regions exist (Mitchell and Finnegan, 2009), e. g. using commercial aircraft with suitable precursors of the aerosol dissolved in the jet fuel create the seeding aerosol in the exhaust plume, or directly injecting the substance into the engine exhaust forming the seeding aerosol in the jets' contrails.

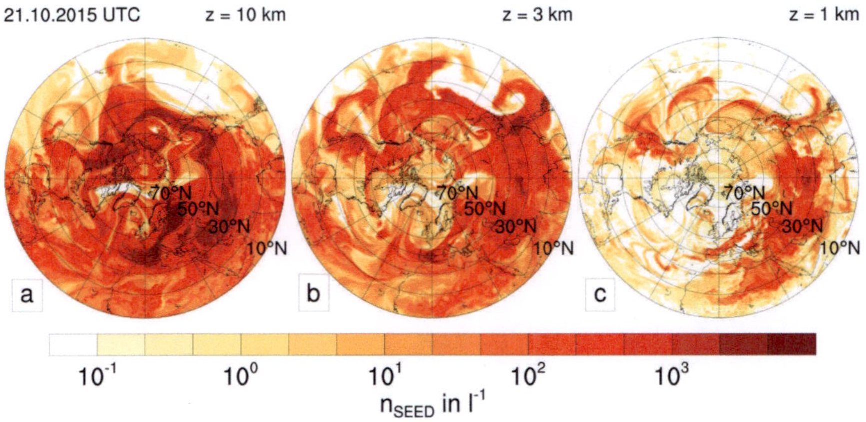

Figure 5.3: Exemplary number concentration of seeding aerosol after 10 days of spin-up for 21 October 2015 in altitudes of 10 km (a), 3 km (b), and 1 km (c).

As already assumed in the preceding study (section 4), such a procedure is mimicked by injecting the aerosol alongside aircraft trajectories at cruise altitude throughout the Northern hemisphere. The trajectories were obtained using traffic way-point information from ADS-B transponders on commercial airliners according to the method described in Gruber et al. (2018). Trajectories sampled over 24 h for a single day are shown in Fig. 5.2. As, for each location and height of the single aircraft, also a time stamp is available, the injection of the seeding aerosol along a certain trajectory is not constant with time, instead, the emissions have diurnal cycles repeated every 24 h.

One example of the resulting distributions of the seeding aerosol after a 10 day spin-up simulation is show in Fig. 5.3 for various altitudes. As already described for the high-resolution process study (section 4.3), the maximum concentration of seeding aerosol is located at the main cruise altitudes, visible as notable maximum concentrations at 10 km altitude (Fig. 5.3a). Despite the absence of sources, i. e. commercial flights over most parts of the Arctic region (Fig. 5.2), long-range transport, mostly due to the strong hibernal jet stream effectively carries the seeding aerosol into the region. In altitudes beneath, the concentrations drop about one order of magnitude (Fig. 5.3b). In the PBL, the aerosol is diluted due to turbulent processes and removed from the atmosphere by deposition and washout.

However, such a widely unbridled spreading of the seeding material results in non-negligible concentrations occurring also far off the target region in low latitudes. Here, the efficacy of seeding cirrus clouds with respect to increase OLR at TOA is much lower than over the hibernal Arctic (Storelvmo et al., 2014). Also, due to already high concentrations of natural aerosol and thus dominance of heterogeneous nucleation in theses areas, additional seeding yield the opposite effect, hence resulting in enhanced values for n_{ICE} and optically thicker clouds. For the other starting dates used, the distribution patterns are fairly similar (not shown).

5.3 Validation

Validation of simulations with a highly complex and detailed modeling system such as ICON-ART, is an important, yet challenging task. Several aspects, for example different processes interact and rely on each other, therefore, it is hardly possible to untangle different impacts; simply neglecting processes causes other ones to react in a non-linear manner. Therefore, the entire setup of the model incorporating all aspects, hence as well the impact of the changed numerical method to calculate the microphysics as the revised calculation of cloud optical properties and the overall model performance are considered here.

In a straightforward manner, only an exemplary validation is presented here. The comparison with both satellite data as well as data gathered by means of aircraft-based remote sensing has yielded adequate agreement to a carefully adjusted high-resolution process study (section 4). Therefore, the reference simulation for the same day (20 January 2016) performed within the scope of the present study, is compared to the same data.

5.3.1 Extinction coefficient

In Fig. 5.4, the retrieval for the extinction coefficient at 0.532 μm from v3 CALIOP Level 2 Profile Cloud product, as described in section 4.2.1, is compared to the extinction coefficient at the nearest ICON wavelength 0.550 μm for the reference simulation starting at 20 January 2016 (*REF06*). Both location and time for comparison are chosen in accordance to the setup for high-resolution study in section 4, thus data sampled along all swaths of CALIPSO at 20 January 201 over the region covered by the domain depicted in Fig. 4.1 are compared.

As for the high-resolution simulation, observation and simulation agree reasonably well. However, also with the coarse resolution of *REF06*, the frequency of occurrence of extinction coefficient larger than 1 km^{-1} at

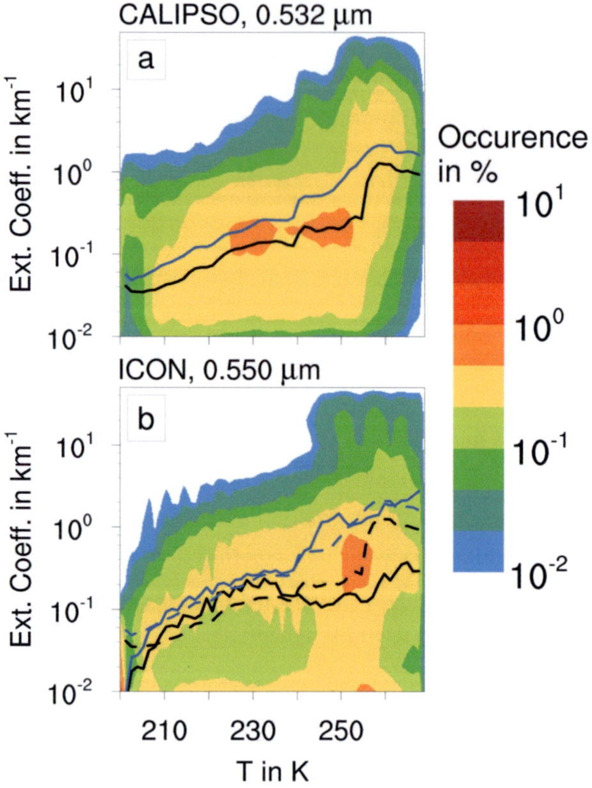

Figure 5.4: Two-dimensional probability distribution of the extinction coefficient with respect to temperature sampled along all tracks of CALIPSO crossing the section of the global simulation area of identical to the domain for the high-resolution simulations (Fig. 4.1) at 20 January 2016 for CALIPSO (a) and simulation *REF06* (b). Black (blue) lines indicate average (median) extinction coefficient in each temperature bin. In (b), average and median from (a) are shown as dashed lines.

temperatures below 230 K, i. e. at higher altitudes is overestimated compared to CALISPO, although the agreement is better than found in before (Fig. 4.5 in section 4.2.1).

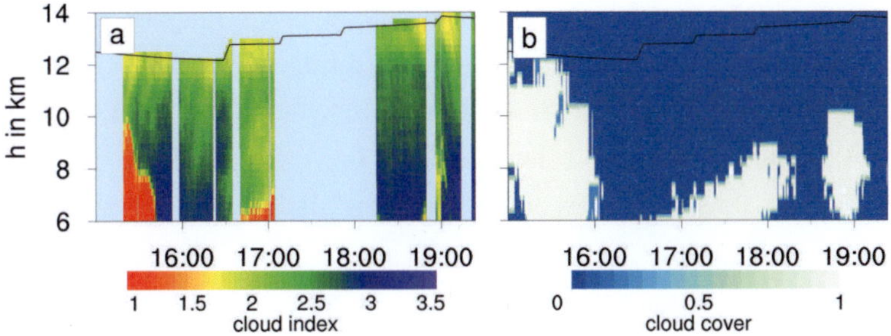

Figure 5.5: (a) Vertical cross-section of cloud index derived from GLORIA measurements during PGS08 on 20 January 2016. Low cloud index values indicate opaque conditions due to cloud particles along the instrument's line of sight. (b) Cloud mask derived from the model simulation *REF06* at the geolocations of the measurement (0 = no cloud / 1 = cloud present). Black line in both panels: HALO flight altitude. Times in UTC.

5.3.2 Supersaturation

Also the comparison with GLORIA retrievals obtained during PGS08 is repeated for simulation *REF06* to assess the distribution of both humidity and supersaturation. In addition, the cloud index derived from GLORIA measurements, i. e. a measure the presence of clouds determined by their opaqueness, is used to validate the simulated spatiotemporal occurrence of cloud patterns. Like for the analysis presented in section 4.2.2, *REF06* starts at 00 UTC, therefore, the simulation data has a lead time of 15 to 19.5 hours, depending on the specific part of the flight considered. Optically thick, and thus opaque, clouds are indicated by cloud index values close to 1, values larger than 4 occur when no clouds are present; values in between indicate either fractional cloud cover or optically thin clouds.

The cloud index along the chosen section of the flight is shown in Fig. 5.5a. Gaps in the data, indicated by grey colors, are mainly due to either a different operation mode, or rather calibration of the instrument. Cloud patterns in the simulations are evaluated along the geolocations of the GLORIA data (Fig. 5.5b); for a comprehensive overview, also shown at those points,

where GLORIA data is missing. The cloud cover ranging from cloud-free conditions (0) to dense clouds (1) is determined by scaling the present number concentration of both frozen and liquid hydrometeors with a particular threshold value, indicating a dense cloud. For the sum of ice crystals and snow flakes, a threshold value of $n_{\mathrm{CRIT}} = 0.1\ \mathrm{l}^{-1}$ is chosen, for cloud droplets, $n_{\mathrm{CRIT}} = 1\ \mathrm{cm}^{-3}$.
At the beginning of the cross-section, the GLORIA data show a dense cloud, i. e. cloud index close to 1, with a cloud top decreasing from about 10 km to below 6 km at 15 : 40 UTC. A similar cloud structure is found in the model data, beginning at about 12 km and decreasing at about 16 : 00 UTC at an altitude of 8 km. While the shape of the cloud top is similar, systematic differences of the cloud top altitudes are probably due to a slightly different horizontal location of the cloud system in the simulation. Another dense cloud below about 7 km to 7.5 km stretching in the GLORIA data from shortly after 16 : 30 UTC to beyond 17 : 00 UTC is reproduced in location and horizontal extent by the model, while the cloud ceiling is slightly underestimated. In the model data, the cloud system arches further up to about 9 km in the measurement gap and declines below 6 km at 18 : 30 UTC. Finally, an optically less dense cloud is found in the GLORIA data in a narrow time interval from about 18 : 45 UTC to 19 : 00 UTC below 9 km. In the same time interval, the forecast data show a similar cloud structure at nearly the same altitudes, but with a slightly larger horizontal extent.
Figure 5.6 shows GLORIA data of water vapor volume mixing ratio and relative humidity (i.e. ice supersaturation) together with the corresponding model data. The GLORIA and ICON-ART data consistently show typical stratospheric water vapor mixing ratios well below 10 ppmv around and above 10 km (Fig. 5.6b and d/f). The simulated increase of water vapor at tropospheric altitudes below 9 km between 15 : 45 UTC to 16 : 30 UTC agrees very well with the observation. Furthermore, the simulated increase of water vapor towards tropospheric values below 10 km after 18 : 15 UTC agrees reasonably well with the GLORIA data.

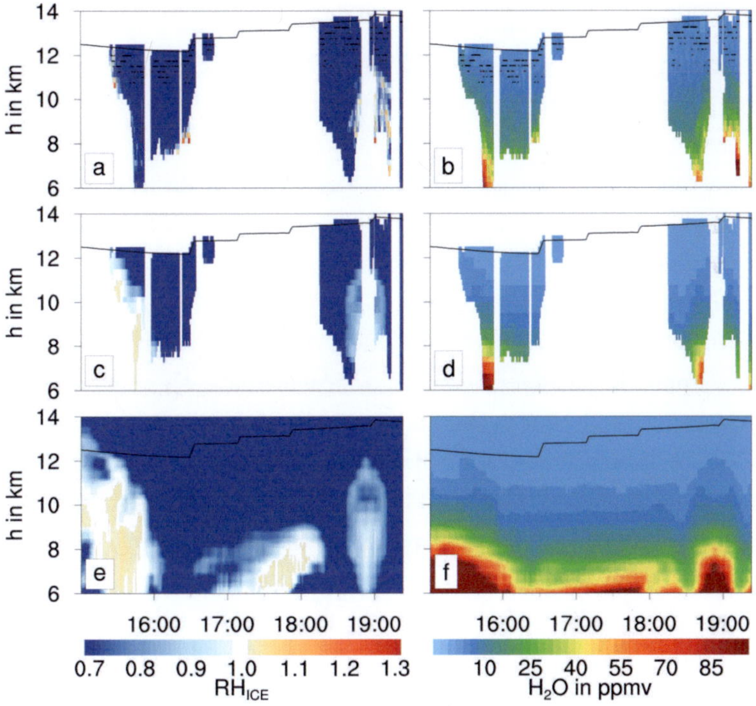

Figure 5.6: Cross sections of relative humidity (i.e. supersaturation versus ice phase) (a) and water vapor volume mixing ratio (b) derived from GLORIA measurements during the HALO flight PGS08 on 20 January 2016. Relative humidity (c) and water vapor volume mixing ratio (d) simulated by the model and filtered for geolocations where GLORIA data are available. Full vertical cross section of the same parameters from the model simulation *REF06* (e,f). Times in UTC.

The comparison of relative humidity derived from the GLORIA and model data is shown in Fig. 5.6a and c/e. Relative humidity well below 1 (blue) is consistently found in regions characterized by low stratospheric water vapor volume mixing ratios (Fig. 5.6b and d/f) and free of clouds (Fig. 5.5). In the close vicinity of clouds seen in the GLORIA data, relative humidity consistently increases towards 1 and higher in the observation. Similarly, the model simulates enhanced relative humidity in regions covered by clouds and their close vicinity. The slightly different positions of cloud systems in observation and forecast are thereby reflected in the relative humidity distributions (e.g. region covered by a cloud and characterized by enhanced supersaturation in the model cross section before 16 UTC).

Overall, the ICON-ART simulation shows a high degree of agreement with the GLORIA observations in context of the forecast period. All tropospheric cloud systems found in the observation can be identified clearly in the simulation. Differences in the locations and fine structures of the clouds are explained to the distinct dynamics inherent to the forecast. While the thermal tropopause is mostly located about 1 km to 2 km lower in the simulation, the observed absolute water vapor volume mixing ratios are reproduced well. In context of the slightly different locations and patterns of cloud systems in observation and simulation, relative humidity simulated by the model is also in agreement with the GLORIA data.

In addition, further comparison with suitable measurement data available, as listed in Tab. 5.1, are shown in appendix E. The flight paths of HALO for PGS10, PGS14, PGS17, and PGS19, limited to the Arctic region, are depicted in Fig. E.1. For those flights, comparison of cloud index determined from the measurement and cloud cover of the corresponding simulations are shown in Fig. E.2, Fig. E.4, Fig. E.6, and Fig. E.8. As discussed above for PGS08 and *REF06*, also for these situations, the model reproduces the observed cloud patterns to a reasonable extent.

Similarly, the comparison of water vapor volume mixing ratio and relative humidity with respect to ice from GLORIA retrievals and model data is in good agreement, as depicted in Fig. E.3, Fig. E.5, Fig. E.7, and Fig. E.9.

5.3.3 Radiation budget for the Arctic winter 2015 / 2016

A common method to verify climate models is the comparison of the simulated radiation balance at the top of the atmosphere averaged over a suitable long period. However, the study presented here is by far no climate simulation in a broader sense, as only about half a year is considered. Nevertheless, due to the long experiment lead time, a semi-annual cycle can be approximated.
For comparison, data is used from the CERES[1] (Wielicki et al., 1996; Gupta et al., 1997) instrument mounted on the Terra and Aqua satellites as part of the NASA[2] EOS[3] to estimate the radiation budget at TOA, i. e. providing long-term measurements of the reflected solar and emitted thermal radiative fluxes at TOA (Wielicki et al., 1996). Pointing nadir, the CERES instrument measures radiance in wide bands for the shortwave part of the spectrum (0.3 – 5 μm), the atmospheric window (8 – 12 μm) and the entire spectrum (0.3 – 200 μm) (Gupta et al., 1997). In this study, the monthly averages of CRERES Energy Balanced and Filled (EBAF) level 3B edition 4.0 product (Loeb et al., 2009) at 1° resolution is used.
The differences between CERES data and results from all ICON-ART simulations carried out to approximate the Arctic winter 2015 / 2016 are depicted for the LW part of the spectrum in Fig. 5.7. For each of the months October 2015 until March 2016, the monthly mean of the CERES data is considered. For the same time period, the ICON-ART simulations cover only 10 episodes of 3 days each. Comparing both data sets directly on a monthly base, i. e. CERES data averaged over one month compared to a 3-day-mean of the ICON-ART simulation, however, shows large deviations,

[1] Clouds and the Earth's Radiant Energy System
[2] National Aeronautics and Space Administration
[3] Earth Observing System

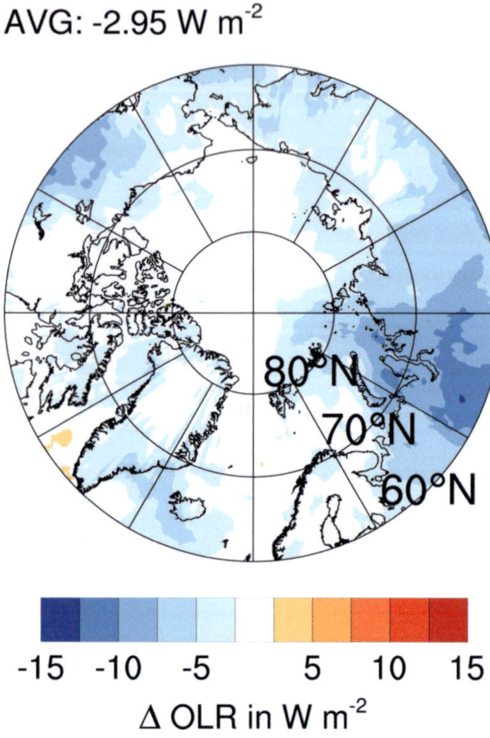

Figure 5.7: Difference in radiation budget at the top of the atmosphere for the period 01 October 2015 until 31 March 2016 between ICON-ART simulations and CERES for outgoing longwave radiation. The mean value for the region north of 60°N is given at the top left.

as especially patterns of cloud cover change a lot during a one-month period, of what the simulation captures only a small fraction. In order to minimize this effect and nonetheless perform a fair comparison, the difference between observation and simulations are indeed calculated for each month separately, but additionally, the resulting monthly differences are averaged over the entire period. Therefore, the results of this analysis cannot serve to validate the accuracy of the model's performance with regard to predicting the exact locations and spatial patterns of clouds; however, systematic biases concerning

the microphysics and subsequently cloud optical properties and the resulting radiative fluxes can be examined.
The deviations between the observed and simulated outgoing LW flux at TOA is depicted in Fig. 5.7. Overall, ICON-ART underestimates the outgoing LW flux, however, differences are small and usually not exceed -5 W m^{-2} to -10 W m^{-2}. Larger deviations are found south of Greenland and Iceland, as well as over Canada. These areas are strongly affected by crossing cyclonic systems, resulting in rather strong temporal variability of cloud cover, by construction only partly captured by the simulations. Over Siberia, the model underestimates the outgoing LW flux by up to -15 W m^{-2} in a semiannual mean. Here, a systematic bias may be present, most likely due to a underestimation in surface temperatures and subsequently too low convective activity. Over the entire domain and time, ICON-ART underestimates the LW balance by approximately -3 W m^{-2}.

5.4 Response to seeding

The Arctic winter 2015 / 2016 is approximated by in total 10 simulations with equally distributed starting dates spread over the entire period, each comprising 72 h. Averages over all of them are considered as mean with respect to the entire period from 01 October 2015 until 31 March 2016.
Each simulation is initialized with distribution of both natural and seeding aerosol that is determined the results of emissions and atmospheric transport during the prior 10 days. Two scenarios are considered, one reference *REF*, i. e. without considering the effect of seeding aerosol on cloud formation and one *SEED* scenario including these processes. However, the different simulations pooled into *SEED* do not interact with each other, thus changes in the atmospheric structure or dynamics caused by modified cloud patterns do not propagate through the entire period. This shortcoming of the approach using time slices instead of one continuous simulation is acquiesced in favor

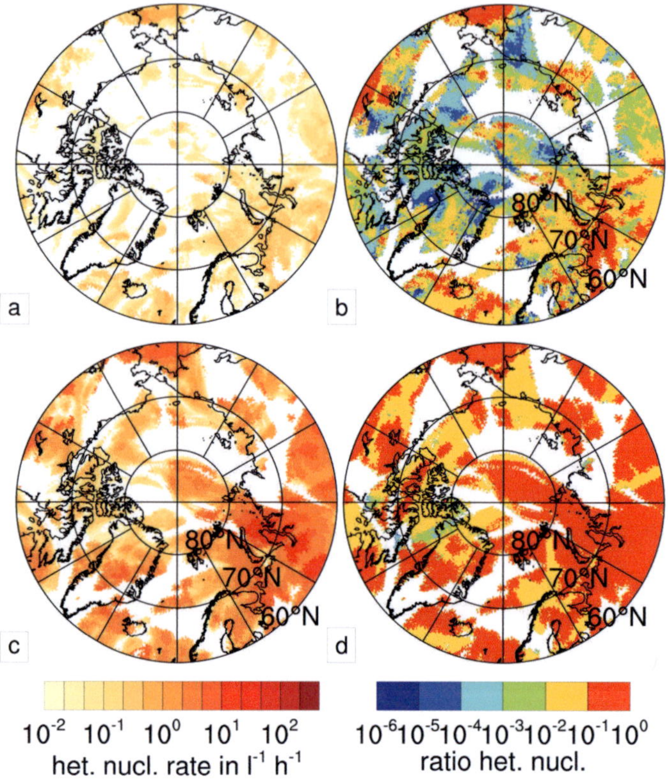

Figure 5.8: Heterogeneous nucleation rate (left) and ratio of heterogeneous to total nucleation rates (right) for *REF* (top) and *SEED* (bottom) averaged over the entire domain and Arctic winter 2015 / 2016.

of having a pair of simulations with and without seeding for meaningful comparisons.

5.4.1 Modification of cloud properties

First, the desired change of nucleation regimes in the homogeneous temperature regime, i. e. $T < 235$ K, is examined. The primary process of a working seeding scenario is the extensive suppression of homogeneous nucleation. Instead of numerous, very small and long-living ice crystals in the cirrus regime with a rather strong net warming effect in the absence of SW radiation, heterogeneous nucleation due to solid aerosol particles forms less but larger and hence faster sedimenting ice crystals. They consume excess water vapor and reduce supersaturation to an extent, the critical threshold for homogeneous nucleation cannot be reached any more. In addition, larger ice crystals have a lower extinction coefficient, are less optically thick and hence contribute less to the net warming effect in the LW spectrum. The heterogeneous nucleation rate occurring with $T < 235$ K averaged over the entire time considered, is shown for *REF* in Fig. 5.8a and for *SEED* in Fig. 5.8c. The ratios of heterogeneous to total nucleation rate for *REF* are depicted for *REF* in Fig. 5.8b and for *SEED* in Fig. 5.8d. Apparently, also without artificial seeding aerosol, heterogeneous nucleation takes place over the entire domain due to the presence of mineral dust particles (Fig. 5.8a). When this effect is dominant, additional seeding aerosol cannot have the desired effect of reducing n_{ICE}, as instead, additional ice crystals are produced. Areas, where this is the case, are found in Fig. 5.8b over Siberia, northern Canada, the North Pacific and south of Iceland. However, over the vast rest of the domain, the contribution of heterogeneous nucleation to cloud formation is negligible. Therefore, in *SEED*, where heterogeneous nucleation strongly enhanced (Fig. 5.8c) due to high concentrations of seeding aerosol in the decisive altitudes (e. g. Fig 5.3a). Consequently, the ratio of heterogeneous to total nucleation reaches values close to 1 in large parts of the region (Fig. 5.8d). These findings hence confirm the results of section 4 for both a longer simulation period as well as for enough cases to reasonably approximate an entire Arctic winter.

Height profiles of several crucial properties of the atmospheric ice phase are averaged over the area north of 60°N and the entire winter are shown in Fig. 5.9. The number concentrations of ice crystals for both scenarios are depicted in Fig. 5.9a. The maximum values for $n_{\mathrm{ICE}} = 110\ \mathrm{l}^{-1}$ in the altitudes where cirrus clouds are found, is less than half of what was found in a high-

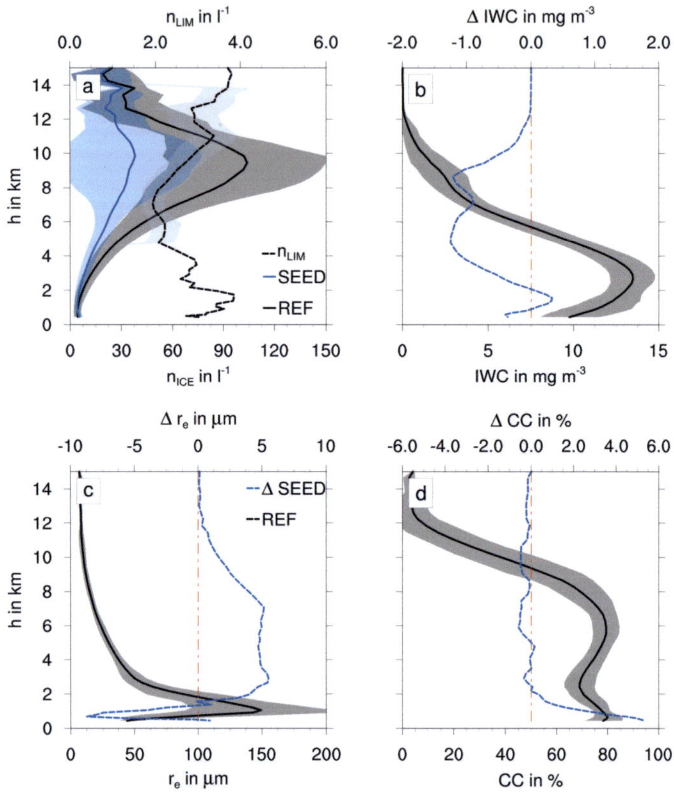

Figure 5.9: Comparison of ice cloud properties averaged over the entire domain and Arctic winter 2015 / 2016 for *REF* (black) and *SEED* (blue). a) n_{ICE}; b) *IWC*; c) r_e; d) cloud cover. Solid lines are absolute values, dotted lines represent differences compared to the reference simulation. Shaded areas indicate standard deviations. Note that in a) only absolute values are depicted, the black dotted line is n_{LIM} for *REF*.

resolution simulation (Fig. 4.15a in section 4). This is to be explained with the impact of a larger horizontal grid spacing reducing the grid-scale vertical velocity. Nevertheless, a key result in section 4 is the relative independence of the reduction of n_{ICE} due to seeding from changes in several environmental conditions such as the properties of sulphate aerosol and especially from the assumed strength of turbulence broadening or narrowing the width of the sub-grid scale distribution of w. These findings can be confirmed, as also here, with *SEED*, n_{ICE} is reduced strongly, with values close the ones found in section 4 (Fig. 4.15a).
In contrast to the strong reduction of n_{ICE}, only minor absolute changes in IWC are found in the cirrus regime with seeding (Fig. 5.9b). In *REF*, IWC is below 5 mg m^{-3} here. Reduction due to seeding in *SEED* about 1 mg m^{-3} corresponds to a reduction of up to 50 %.
Also the effective radius of ice crystals increase only marginally with seeding (Fig. 5.9c). This is due to the assumption made concerning the shape of the ice crystals being hexagonal columns. For them, a doubling of the mean mass, i. e. IWC/n_{ICE}, leads to an increase in r_{e} of only about $1/4$.
One important effect of seeding cirrus clouds is thought to be a reduction in high-level cloud cover caused by increased sedimentation rates of the larger ice crystal formed on the seeding aerosol particles (Mitchell and Finnegan, 2009; Storelvmo et al., 2013; Gasparini et al., 2017). However, in the presented study, this does not happen. In a former study (section 4), where also hardly any reduction in cirrus cloud cover was found, this was attributed to the short integration time, in which there was not enough time for the ice crystals to sediment. Furthermore, due to the limited area of the simulations and the need of updating the boundaries with undisturbed values for cloud properties and humidity, no large scale impact of seeding was found. However, in the presented study, enough time for ice crystals to be removed by sedimentation is available. In addition, also the impact of updating the boundary is omitted. Nonetheless, in Fig. 5.9d, hardly any reduction in high-level cloud cover is

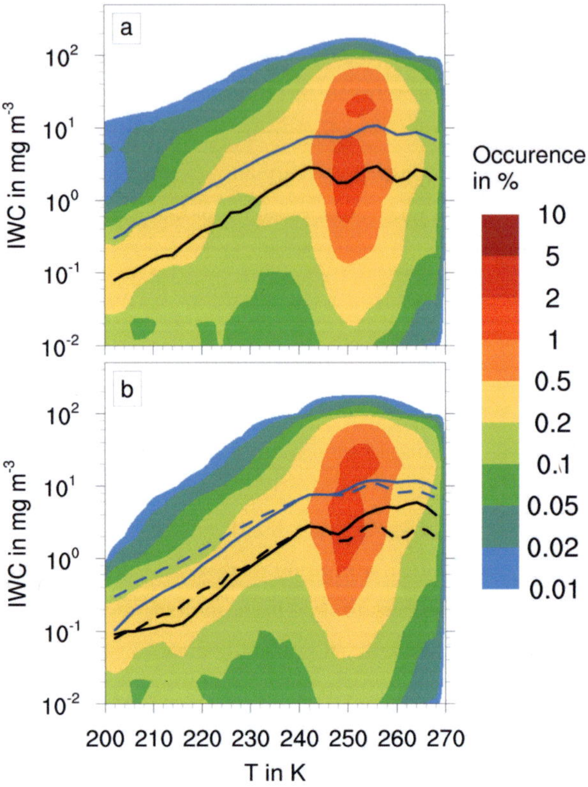

Figure 5.10: Two-dimensional probability distribution of the *IWC* with respect to temperature averaged over the entire domain and Arctic winter 2015 / 2016 for *REF* (a) and *SEED* (b). Black (blue) lines indicate average (median) *IWC* in each temperature bin. In (b), average and median from (a) are shown as dashed lines.

found due to seeding. This is mainly due to the definition of cloud cover, here defined as n_{ICE} exceeding a certain threshold.

For a more comprehensive investigation, probability distributions of *IWC* for temperature bins of 2 K sampled over the entire domain an simulation time are shown in for *REF* in Fig. 5.10a, and for *SEED* in Fig. 5.10b. Obviously, *IWC* is largest and occurs most frequently with higher temperatures, i. e. at

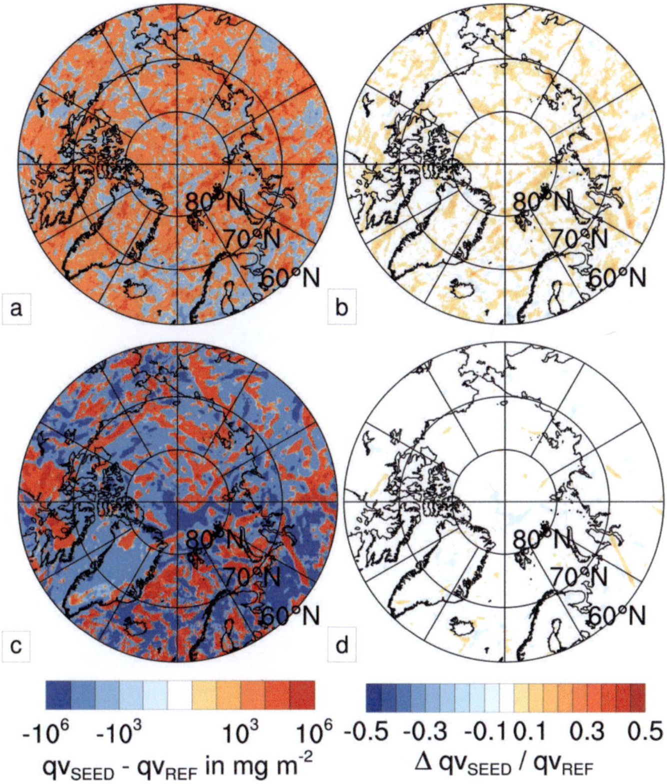

Figure 5.11: Difference in column integrated water vapor content between *SEED* and *REF* in regions with $T < 235$ K (top) and 235 K $< T <$ 273 K (bottom) averaged averaged over the entire domain and Arctic winter 2015 / 2016. Left: absolute difference, right: relative difference.

lower altitudes for both scenarios. However, at temperatures lower than 230 K, the frequency of occurrence for large values of *IWC* between 1 mg m^{-3} and 10 mg m^{-3} is reduced and slightly shifted to values lower than 1 mg m^{-3}. This indicates, that in this study, seeding cannot dissolve cirrus clouds entirely, but at least a certain reduction in their *IWC* and hence their optical depth is achieved. In association with a slight increase in r_e, this can impact the radiative fluxes.

Although there is an impact of seeding on cirrus clouds, the environmental conditions are modified only marginally. Especially, stability within the altitudes, where cirrus clouds are present, hardly changes (not shown). Also, relative humidity is on average not modified in a decisive, statistically significant manner. However, over certain areas, atmospheric water vapor content changes due to seeding. In the top row of Fig. 5.11, the difference in column integrated water vapor content between *SEED* and *REF* in regions with $T < 235$ K averaged over the entire simulation time is depicted as absolute (Fig. 5.11a) and relative difference (Fig. 5.11b).
The same quantities but for the temperature ranging from 273 K until 235 K, are shown in the bottom row of Fig. 5.11.
Collocated to the areas, where n_{ICE} is reduced within the homogeneous nucleation regime (not shown), a certain increase of up to 10^3 mg m^{-2}, corresponding to about 10 % in water vapor content is found (Fig. 5.11a and Fig. 5.11b). The absolute differences are larger with temperatures above 235 K in Fig. 5.11c, although the relative impact is smaller (Fig. 5.11d). This is due to the generally larger absolute water vapor content within those lower, hence warmer, altitudes. Especially within the temperature regime below 235 K, homogeneous nucleation is extensively suppressed when seeding aerosol is present, as the latter acts as IN. Apparently, consumption of the equal amount of excess water vapor content by depositional growth of the ice crystals formed by heterogeneous nucleation on the seeding aerosol takes about 100 times longer than by reducing supersaturation during homogeneous freezing (Köhler and Seifert, 2015). These results, at least qualitatively, confirm the findings of Gasparini et al. (2017). Here, applying a similar seeding scenario, in the annual mean, reductions in relative humidity between 1 and 5 % in the polar upper troposphere (their Fig. 5f) were found. However, as water vapor is among the most effective greenhouse gases, increasing concentrations counteract the cooling achieved by thinning cirrus clouds.

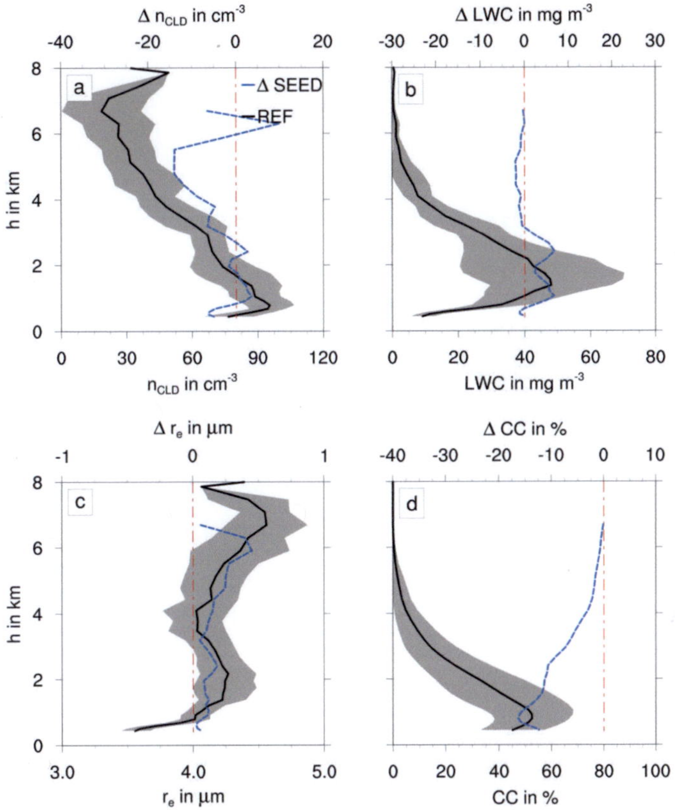

Figure 5.12: Comparison of liquid cloud properties averaged over the entire domain and Arctic winter 2015 / 2016 for *REF* (black) and *SEED* (blue). a) n_{CLD}; b) LWC; c) r_{e}; d) cloud cover. Solid lines are absolute values, dotted lines represent differences compared to the reference simulation. Shaded areas indicate standard deviations.

In the altitudes beneath the cirrus regime, i. e. between about 3 km to 8 km, the reduction in n_{ICE} is much smaller than above (Fig. 5.9a), although the reduction in *IWC* is relatively constant with height at about 1 mg m^{-3} (Fig. 5.9b). As mean ice crystal masses and hence the effective radius is larger in those altitudes, also the increase due to seeding increases to about 5 μm between 3 km to 7 km (Fig. 5.9c); corresponding to an increase of about 10 % to 25 %. However, also in these altitudes, with less than 1 %, hardly any change in cloud cover is found (Fig. 5.9d).

In contrast to the response of the ice phase, the liquid phase is strongly modified when introducing seeding aerosol. Fig. 5.12a, the averaged vertical profile of cloud droplets number concentration (n_{CLD}) is shown. Here, a slight decrease of about 10 % is found relatively constant with height. In contrast, *LWC* remains merely unchanged (Fig. 5.12b). Consequently, the effective radius increases slightly (Fig. 5.12c). On average, cloud droplets do not change much, neither with respect to size, nor with respect to number concentration.

However, as already seen in section 4, the largest response to seeding within the liquid phase is a distinct reduction in low level cloud cover of up to -20 %, i. e. especially in altitudes below 3 km, cloud cover is nearly halved. Accompanied by only a small increase of cloud cover in the ice phase (Fig. 5.9), consumption of excess water vapor by increased numbers of ice crystals and hence reduced supersaturation available for activation of CCN is only a minor effect leading to reduced liquid cloud cover.

Indeed, additional ice crystals form due to the presence of seeding aerosol within those altitudes. The larger effect of those ice crystals however, are enhanced riming rates, or Wegener-Bergeron-Findeisen process (e. g. Fan et al., 2011), i. e. collisions of cloud droplets and ice crystals and subsequent sticking. The resulting hydrometeors are sorted into the graupel class.

With seeding, especially within the cirrus regime, n_{ICE} is reduced strongly, throughout the free troposphere, also a remarkable decrease in *IWC* is found, despite grid-scale cloud cover remains merely constant. Additionally, cloud cover of liquid clouds is reduced strongly within the PBL. Consequently, the

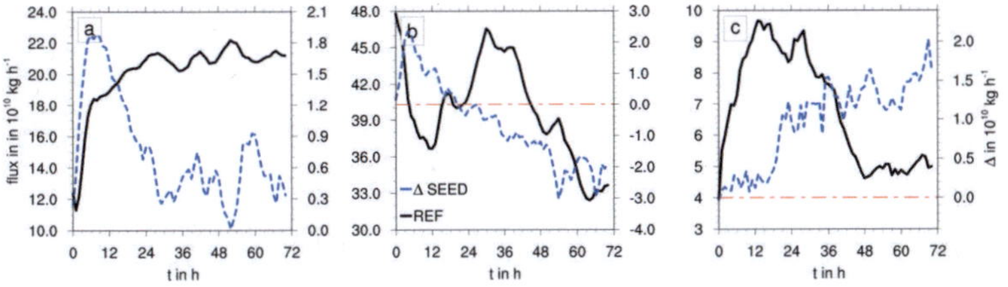

Figure 5.13: Comparison of precipitation rates integrated over the entire domain and Arctic winter 2015 / 2016 for ice crystals (a), snow (b), and graupel (c). Absolute values in black for *REF*; difference between *SEED* and *REF* in blue.

question after the sink of the hydrometeors arises. To elucidate this issue, the temporal evolution of precipitation rates for ice crystals, snow, and graupel averaged over the entire domain and the members of both *REF* and *SEED* are show in Fig. 5.13.

During the first 6 h, the rates in *REF* increase strongly for ice and graupel, and decrease for snow. This is most likely to be explained by the adjustment from the initialization towards a balanced state determined by the chosen microphysical scheme. Also, even the larger hydrometeors require a certain time until forming and sedimenting down to the ground.

However, also distinct differences between *REF* and *SEED* are found in the first 24 h. During that time, the larger ice crystals, snow flakes formed by collisions of ice crystals, and graupel, mainly formed from riming cloud droplets onto both snow and ice particles, reach the ground.

After about 24 h, the precipitation rate of ice crystals in Fig. 5.13a remains on a relatively constant value for both *REF* and *SEED*, with larger values in *SEED*, the latter obviously caused by increased sedimentation rates from the cirrus layer and beyond.

The mass flux of snow reaching the ground monotonously decreases with time *REF* and does so even more pronounced in *SEED*; the latter being caused by enhanced riming rates prohibiting both snow formation and even convert snow into graupel, hence reducing mass concentrations of snow.

In contrast, the difference between *SEED* and *REF* in mass flux of graupel reaching the surface increases with time (Fig. 5.13c), although the percentage of grid cells covered by liquid clouds is strongly reduced within the PBL in *SEED* (Fig. 5.12d). This hints at a continuous source of both moisture and liquid clouds contributing to graupel formation transported into the domain from regions further south.
In total, i. e. over the entire integration time of each simulation, the mass flux of ice crystals is increased by about 4 %, the precipitation rate for graupel is enhanced by nearly 15 %.

5.4.2 Impact on radiation

The impact of injecting artificial aerosol into northern Polar regions' the upper troposphere and the consequent changes of cirrus and mixed-phase clouds is investigated over in total 10 different cases spread in the period October 2015 until March 2016, hence approximating the Arctic winter.
Cloud coverage in high altitudes hardly changes due to seeding. Nevertheless, cirrus clouds are found to get thinner, i. e. ice crystal number concentrations are reduced. This leads to a small increase in the effective radius. Jointly with a decrease in IWC, the warming effect of these clouds is reduced, as their optical depth decreases. Due to the presence of seeding aerosol also in lower altitudes, this response to seeding occurs also there, although less distinct. However, a small increase in water vapor content, at least partly, counteracts this effect.
In addition, mixed-phase clouds are found to glaciate leading, especially within the PBL, to a strong decrease in coverage of liquid clouds.
Although the changes are each rather small, however, when superimposed, the effect on the radiative fluxes is large. In Fig. 5.14, the temporal evolution of the spatial mean LW fluxes during the Arctic winter 2015 / 2016 are shown for both the Earth's surface (Fig. 5.14a) and at TOA (Fig. 5.14b). The spatial mean LW radiative flux at TOA follows the solar circle, hence decreases until midwinter and increases afterwards. For the same area and time, the mean for

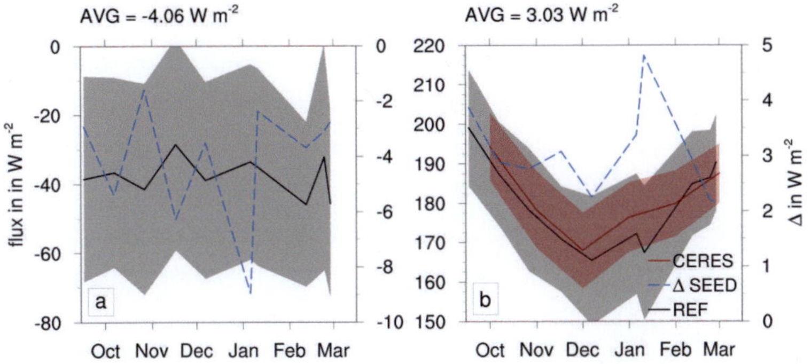

Figure 5.14: Temporal evolution of spatial mean of LW radiative fluxes for Arctic winter 2015 / 2016 at the surface (a) and TOA (b). Solid black (red) lines and shades indicate absolute spatial mean and standard deviations with respect to spatial mean for *REF* (CERES data); dashed blue lines are differences between *SEED* and *REF* with mean values for the entire period given top right of each figure.

the same area and time obtained from CERES data (red line in Fig. 5.14b) agrees fairly well with the simulations, despite a systematic underestimation of 1 to 5 W m^{-2}, depending on the particular situation (see also section 5.3.3). A positive anomaly of the LW flux at TOA, hence increasing OLR, generates radiative cooling, as energy contained within the atmosphere is transported into space. This effect, caused by thinning cirrus clouds, occurs during the entire Arctic winter, with a magnitude between 1 to 5 W m^{-2}. However, the efficacy of seeding found here is somewhat smaller than the response found in a higher resolved, semi-idealized study (section 4.3.5 in section 4). The single case (20 January 2016) that study focuses on, is the one, where seeding has the largest effect during the entire period considered. Due to the lower values of n_{ICE} in this study caused by smaller sub-grid-scale fluctuations of the vertical velocity due to the larger grid spacing, the absolute reduction caused by seeding is decreasing, hence also difference in the radiative fluxes are smaller. Also, the domain considered here is by far larger than the one used in section 4 covering areas, where seeding is not as effective, e. g. over the north of the Asian continent. However, the efficacy of seeding found here is somewhat smaller than the response found in a higher resolved, semi-idealized study (section 4.3.5 in section 4). The single case (20 January 2016) that study focuses on, is the one, where seeding has the largest effect during the entire period considered. Due to the lower values of n_{ICE} in this study caused by smaller sub-grid-scale fluctuations of the vertical velocity due to the larger grid spacing, the absolute reduction caused by seeding is decreasing, hence also difference in the radiative fluxes are smaller. Also, the domain considered here is by far larger than the one used in section 4 covering areas, where seeding is not as effective, e. g. over the north of the Asian continent.

On average, in the seeding scenario considered here, OLR is increased by about 3 W m^{-2}, hence is able to outweigh the long-term global mean impact of a doubling of CO_2 of about 1.4 to 4.5 W m^{-2} (Holland and Bitz, 2003). However, such a comparison is joint with several caveats, mainly addressing the spatial and temporal limitation accompanied with the focus on a single Arctic winter. Nevertheless, the effect of seeding on OLR found here agrees fairly well with former studies, e. g. Gasparini et al. (2017) found a maximum average reduction in OLR of about -3 W m^{-2} for long-term simulations assuming a similar seeding scenario. The presented study underestimates reduction in OLR clearly compared to a rather idealized climate study with a global mean decrease of up to -7 W m^{-2} (Storelvmo et al., 2013). In contrast, assuming a more realistic, spatially non-uniform seeding scenario, temporal means within the northern Polar circle, values of 1 to 6 W m^{-2} are found (Storelvmo and Herger, 2014); agreeing much better to the results presented here.

In addition to the primary effect of seeding on OLR, also the secondary effect of increased upward LW fluxes at the surface, described in section 4, is found here (Fig. 5.14a). During winter, the Arctic surface is constantly cooling, indicated by negative values for the fluxes in *REF*. With seeding, the insulating layer consisting of mixed-phase clouds, partly is removed, or at least reduced, hence allowing more LW radiation escape from the surface and the adjacent atmospheric layers. Compared to section 4, the efficacy of seeding is with an

increase of 9.5 W m^{-2} more than doubled for 20 January 2016. In general, the increase in surface LW fluxes is larger than the OLR during the entire winter. However, several caveats are accompanied with this finding. Sea surface temperatures are fixed to the climatological mean for each case and thus do not change due to radiative cooling. Both ice as well as soil temperatures can adjust to the radiative forcing, but this occurs on time scales larger than the integration time used here. Therefore, although the adjacent atmospheric layers experience a certain cooling, temperatures at the surface remain merely unchanged. The temperature difference increases causing an increase in upward LW fluxes, leading to an overestimation of this effect. Nonetheless, at least qualitatively, the results found in section 4 can be confirmed here.
Besides the LW fluxes, also the SW budget is altered by seeding. As especially cirrus clouds have a cooling impact in the SW regime, this can reduce the overall efficacy of seeding cirrus clouds. However, during most of the time, this effect is negligible due to lacking incoming solar radiation over a large part of the domain. During the beginning and the end of the Polar night, for the cases in October and March, respectively, the reduction in net SW fluxes at TOA does not exceed 1 W m^{-2}; with an overall average of 0.34 W m^{-2} (not shown). Similarly, also the impact of the SW fluxes at the surface is of minor extent with 0.24 W m^{-2} on average.

5.5 Summary

In section 4, a high-resolved process study on thinning of cirrus clouds during Arctic winter revealed a certain potential efficacy of this climate engineering method. In continuation, this section is dedicated to gaining a more comprehensive view on this topic with simulations of the entire Arctic winter 2015 / 2016 using a time slice method. For this, the atmospheric modeling system ICON-ART coupled with a comprehensive and detailed microphysical scheme. Besides a prognostic treatment of the natural aerosol, also the artificial seeding aerosol undergoes the full atmospheric life cycle after being emitted into the upper troposphere alongside commercial aircraft trajectories at cruise altitude.

Validation of simulated cloud properties and environmental atmospheric conditions with data obtained by both the CALIPSO satellite and GLORIA mounted on the HALO aircraft during the POLSTRACC campaign yield reasonable agreement.

In addition, the simulated longwave radiation budget at the top of the atmosphere, as approximated mean throughout the entire period is compared with CERES data, showing a well reproduced spatial pattern although the model systematically underestimates the outgoing longwave fluxes by about 3 W m^{-2}.

The main effect of injecting seeding aerosol into the Polar upper troposphere is a strong reduction in cirrus ice crystal number concentrations. In addition, the ice water content decreases, leading to an, on average, larger effective radius. However, an effective reduction of high-level cloud cover is not found. Also throughout the free troposphere, a remarkable decrease in ice water content is found, despite grid-scale cloud cover remains merely constant.

As consumption of excess water vapor by depositional growth of the ice crystals formed by heterogeneous nucleation on the seeding aerosol is slower than reducing supersaturation during homogeneous freezing, a small increase in atmospheric water vapor content is found confirming the results of a former study (Gasparini et al., 2017).

Additionally, a secondary effect is found, as cloud cover of liquid clouds is reduced strongly by up to -20 %. Both ice crystals sedimented from the upper troposphere as well as newly generated ice crystals from sedimented seeding aerosol lead to glaciation of the liquid phase via enhanced riming rates of cloud droplets, leading to precipitation of graupel. These results agree very well with the findings of a preceding, cloud-resolving process study (section 4).

Thinning of both high- and low-level clouds lead to increasing outgoing longwave radiative fluxes at the top of the atmosphere of about 3 W m^{-2}, hence is able to outweigh the long-term global mean impact of a doubling of CO_2.

This effect occurs during the entire Arctic winter with a magnitude between 1 to 5 W m^{-2} and agrees fairly well with former studies (e. g. Gasparini et al., 2017; Storelvmo and Herger, 2014).
Also a strong average increase of upward longwave radiation at the surface of about 4 W m^{-2} is found, as the insulating layer of low-level mixed-phase clouds is partly removed, hence allowing more longwave radiation escape from the surface and the adjacent atmospheric layers. However, as in this study, the Earth's surface cannot adjust to radiative forcing due to prescribed and constant sea surface temperatures, the effect might be overestimated.
The results presented here are to be interpreted considering several caveats. Due to the approach of using time slices, no large-scale impact of modified cloud properties and radiative fluxes on global dynamical patterns can be investigated, as most likely, such structures change on longer timescales than the ones considered here. However, as large-scale dynamical pattern may adjust to the different forcing, this process might compensate for the effect of seeding after a certain time. In addition, the processes leading to the formation of cirrus clouds, as being the particular target of this climate engineering approach, is afflicted by uncertainties and lack of understanding.

6 Conclusions

6.1 Contrails and their impact on shortwave radiation and photovoltaic power production

In the first part of this thesis, a fully online coupled description of the life cycle of contrails and contrail cirrus and their interaction with the environmental water vapor budget, the natural cirrus clouds, and the radiative fluxes was developed for being used with the regional atmospheric model COSMO-ART. Conducting a case study, the following questions are addressed:

- To what extent differs aircraft induced cloudiness from natural cirrus clouds?
- How large is the impact of contrails and contrail cirrus on the radiative fluxes, especially with regard to photovoltaic power production?
- To what extent can ice crystal loss in young contrails determine contrail-cirrus properties?

To what extent differs aircraft induced cloudiness from natural cirrus clouds?

The formation of contrails is based on processes entirely different from ice nucleation on solid or liquid aerosol particles due to fluctuations in vertical velocity or humidity. Consistent to former findings obtained from in-situ measurements (e. g. Febvre et al., 2009; Voigt et al., 2010; Schumann and Heymsfield, 2017), satellite data (e. g. Minnis et al., 1998; Duda et al., 2004) and numerical simulations (e. g. Schumann et al., 2015; Bock and

Burkhardt, 2016b; Caiazzo et al., 2017), contrails are clearly distinguishable from natural cirrus clouds especially by their much higher ice crystal number concentrations partly exceeding 100 cm^{-3}. After transition into contrail-cirrus, the original line-shaped appearance is blurred; hence a visible distinction compared to natural cirrus becomes difficult. However, also in this stage, number concentrations are still considerably higher compared to natural cirrus. High ice crystal number concentrations limit depositional growth of individual ice crystals, therefore, both mean ice crystal sizes as well as the effective radius of aircraft induced cloudiness is small. This in turn affects the optical properties, leading to remarkable values for both the extinction coefficient as well as the optical depth.

How large is the impact of contrails and contrail cirrus on the radiative fluxes, especially with regard to photovoltaic power production?

Due to their smaller effective radius, hence large extinction coefficients, and consequentially large optical depth, contrails and contrail cirrus potentially have a larger impact on radiative fluxes than natural cirrus clouds. However, young contrails that may have large optical depths, are of small horizontal extent, their absolute effect hence is of minor importance. In contrast, contrail-cirrus can cover large areas, as seen for the case study conducted within this study. For a situation where high-level cloudiness is distinctively governed by aircraft induced cloudiness, the incoming SW radiation can be reduced by about 5 to 10 %. The accompanied loss of PV power production then amounts up to 10 %. However, these numbers have to be put into perspective, as such cases are not too common. Nevertheless, as this phenomenon is not considered in operational weather forecast at all, even a couple of contrail-governed days may cause a distinct economic loss.

To what extent can ice crystal loss in young contrails determine contrail-cirrus?

Contrail formation takes place under extreme thermodynamical and dynamical conditions within the aircraft wake vortices on scales of both time and space too small to be explicitly resolved with the chosen model system. Whereas the number of ice crystals initially forming in the jet exhaust is a well-known quantity, the impact of the early phase turbulence forcing ice crystals to evaporate or being spinned out of the vortices, is much harder to determine. This process is found to have a large and long-lasting impact on contrail ice crystal properties even affecting the properties of contrail cirrus (Unterstrasser, 2014). The results of this study underline the importance of considering this process, as its negligence alters the reduction in incoming SW radiation by up to one third, as contrail ice crystal number concentrations remain very high resulting in optically thick contrail-cirrus.

6.2 Climate engineering by Arctic winter cirrus thinning

In the second part of this thesis, detailed analyses of the impact of a hypothetical injection of aerosol particles into the upper troposphere during Arctic winter were conducted using the online coupled atmospheric modeling system ICON-ART. At this point, a conclusive summary is given.
The examination of the topic was performed regarding to following key questions:

- How are the natural aerosol particles distributed over the Arctic?
- How do cirrus clouds form during the Arctic winter, when considering a realistic background aerosol loading?

- How is the life cycle of the seeding aerosol in the atmosphere including emissions, transport, sedimentation, and ice nucleation in cirrus- and mixed-phase clouds?
- Are there side effects, especially with regard to mixed-phase cloud properties?
- Using suitable methods, can former findings confirmed?
- What are the main uncertainties regarding this method?

How are the natural aerosol particles distributed over the Arctic?

The dominating aerosol species for heterogeneous ice nucleation is mineral dust (Phillips et al., 2008). Although hardly any sources for mineral dust are located within the northern Polar circle, convection and atmospheric currents transport the aerosol particles from there source regions, mainly the large desert areas, to higher latitudes. Partly, number concentrations exceed 100 l^{-1} in the upper troposphere. Also at lower altitudes, mineral dust is present, although in much lower concentrations.

How do cirrus clouds form during the Arctic winter, when considering a realistic background aerosol loading?

Cirrus thinning only can yield the desired effect, when cirrus clouds during Arctic winter mainly form via homogeneous nucleation that can be inhibited by artificially providing suitable aerosol particles. In a semi-idealized study with a realistic vertical profile, constant in time, homogeneous nucleation is by far the prevailing source of ice crystals in cirrus clouds. However, regarding the entire Arctic winter, several large areas dominated by heterogeneous nucleation become apparent. Simple assumptions in long-term simulations, e. g. prescribing a certain ratio of homogeneous to heterogeneous nucleation might therefore be unfavorable.

Apart from this, the parameterization of ice crystal formation in numerical models is to be regarded with certain reservations, not only concerning the description of the physical process itself, but especially with respect to systematic biases caused by the limited resolution of the small-scale fluctuations in vertical velocity and humidity that are decisive for triggering the nucleation process. They become both smaller and more inaccurately parameterized with decreasing horizontal resolution.

Hence, further research on cirrus cloud formation is essentially comprising laboratory experiments on the nucleation process, observations of both aerosol and cirrus cloud properties in the upper troposphere from space and in-situ, and modeling studies to gain more profound knowledge about cirrus cloud properties and occurrence, and their radiative impact. This holds not only with regard to climate engineering studies but, is valid for all kind of numerical studies with cirrus clouds involved. Hence, further research on cirrus cloud formation is essentially comprising laboratory experiments on the nucleation process, observations of both aerosol and cirrus cloud properties in the upper troposphere from space and in-situ, and modeling studies to gain more profound knowledge about cirrus cloud properties and occurrence, and their radiative impact. This holds not only with regard to climate engineering studies but, is valid for all kind of numerical studies with cirrus clouds involved.

How is the life cycle of the seeding aerosol in the atmosphere including emissions, transport, sedimentation, and ice nucleation in cirrus- and mixed-phase clouds?

Limiting the cloud modification on the hibernal Arctic region has turned out to be most promising, as here, due to lack of incoming solar radiation, cirrus clouds have only a warming effect and removal or thinning them yield the maximum possible cooling effect (Storelvmo et al., 2014; Gasparini et al., 2017).

In a thought experiment, commercial aircraft have been proposed, as possible means of emission (Mitchell and Finnegan, 2009). In the presented study, this idea is mimicked by assuming injection of the seeding material along commercial flight trajectories into the atmosphere and subsequent dispersion into the Arctic region.

Despite the utterly absence of commercial aircraft routes over the northernmost parts of the Arctic, seeding aerosol emitted into the upper troposphere of the northern hemisphere, enormous concentrations can build up in the decisive regions within a few days. However, also at lower altitudes, seeding aerosol is present due to sedimentation and turbulent processes.

Largely independent from background conditions and model resolution, the main effect of injecting artificial seeding aerosol into the upper troposphere is a strong suppression of homogeneous nucleation and thus a strong reduction of ice crystal number concentrations down to about 20 % compared to pristine situations. In contrast to former studies (e. g. Storelvmo et al., 2013; Storelvmo and Herger, 2014; Gasparini et al., 2017), high-level cloud cover is not reduced. However, accompanied with a moderate shift to smaller values of the ice water content, a small increase in effective radius is found, both lead to optically thinner cirrus clouds confirming the results of studies applying GCMs (Storelvmo et al., 2013, 2014; Storelvmo and Herger, 2014; Gasparini et al., 2017).

Besides sedimenting ice crystals formed on the seeding aerosol to lower altitudes, the seeding aerosol does so as well, acting as INP also here.

Thinning of cirrus clouds, on average, may increase outgoing longwave radiation by up to 3 W m^{-2}; hence even outweighing the long-term global mean impact of a doubling of CO_2. Although such comparisons have to be regarded with several caveats, the obtained magnitude agrees fairly well with former studies (e. g. Gasparini et al., 2017; Storelvmo and Herger, 2014).

Sensitivity studies show that these findings hold for a large range of possible seeding aerosol sizes and concentrations; over-seeding instead seems to occur only with unrealistically high concentrations. This is contradictory to former studies finding over-seeding already with relatively low seeding aerosol concentrations (e. g. Storelvmo et al., 2013; Gasparini and Lohmann, 2016).

Are there side effects, especially with regard to mixed-phase cloud properties?

The presence of artificial ice crystals within the mixed-phase clouds lead glaciation and consequentially to a strong reduction in liquid cloud cover of up to 20 %, mainly, due to cloud droplets riming onto the ice crystals and forming graupel. Removal of this insulating mixed-phase cloud layer causes a strong increase of upward longwave radiation at the surface of about 4 W m^{-2}. This may amplify the effect of thinning cirrus clouds, as additional cooling of the surface is achieved.

What are the main uncertainties regarding this method?

Several issues have already been mentioned above. Apart from them, the chosen model setup does not allow strong adjustments of large-scale dynamics to the changed forcing. Changing dynamical patterns will most likely have an impact not limited to the Arctic. Former studies found certain changing patterns e. g. regarding the global distribution and amount of precipitation (e. g. Muri et al., 2014; Storelvmo et al., 2014).
Also, as the Earth's surface cannot adapt to the cooling in a reasonable manner, especially the cooling effect in low altitudes may be overestimated.
Concerning the special case of cirrus thinning, however, further long-term simulations with a model system comprising, besides a comprehensive microphysical scheme, also a prognostic treatment of the life cycle of both natural and seeding aerosol, are desirable to confirm the results obtained here.
Nonetheless, this study lines up next to several others performed during the last years demonstrating that thinning cirrus clouds might be a promising approach to cool the Arctic regions (e. g. Storelvmo et al., 2013, 2014; Muri et al., 2014; Kristjánsson et al., 2015). However, the statement given by Mitchell et al.

(2011) is still valid: *it is absolutely critical to understand that this is not a 'silver bullet' for the global warming problem.* Even if, hypothetically, such means of cloud modification were inducted and indeed would yield the desired effect, however, global warming could neither be stopped nor even reverted, but only decelerated at best. Albeit cirrus thinning might lessen the effect for a certain time, it is unable to tackle the origin of global warming, i. e. increasing greenhouse gas concentrations.

6.3 Outlook

In this thesis, two examples of the anthropogenic impact on the upper tropospheric clouds and the Earth's radiation budget were examined.
One key finding is that, at least on special occasion, aircraft induced clouds have a decisive impact on the radiative fluxes, as producing additional high-level cloud cover. These artificial clouds are somehow similar to cirrus clouds, i. e. cause a certain cooling in the SW spectrum and have warming effect in LW part of the spectrum. Due to their special properties, both effects are assumingly larger than for natural cirrus. However, compared to other forcing agents, this impact is considered small but inflicted with large uncertainty (Myhre et al., 2013).
It hence is desirable, to investigate the global impact of contrails on natural cirrus clouds using an online coupled description, as presented in this thesis, in a high- and thus cloud-resolving modeling system like ICON-ART to confirm findings of climate models. In addition, it is crucial considering besides only contrail formation also the composition of aircraft exhaust, as here, effective INP like soot and metallic particles are injected with considerably strong rates into rather pristine altitudes, where they instantly can impact cloud formation. In addition, also several gases with a high impact on the radiative fluxes are emitted alongside.
With respect to climate warming, removal of high-level cloudiness, at least partly, is preferable. Former studies pointed at the possibility to reduce

occurrence of contrails by avoiding flying through supersaturated areas either by choosing different altitudes (e. g. Mannstein et al., 2005; Deuber et al., 2013), or vary the routes (e. g. Soler et al., 2014; Zou et al., 2016). Furthermore, it might be possible to avoid contrail formation by either reducing air traffic at all, or using other types of fuel for combustion reducing number concentrations of contrail ice crystals and hence reducing their optical depth (Irvine and Shine, 2015; Caiazzo et al., 2017; Moore et al., 2017).

Comprehensively quantifying the impact of aviation in such a way may lead to conclusions of either finding other ways to decelerate global warming, and hence a reduced necessity of climate engineering, or rather reduce the extent of its required deployment. Furthermore, it might be decisive for every means of emitting the seeding material involving aircraft, to quantify the impact of contrails produced alongside.

Finally, such findings can be used for conducting long-term climate studies in order to warrant a most comprehensive description of the complex interactions between those aspects and processes.

Besides all these reflections, another key result is the affirmation that, considering all the uncertainties involved, thinning cirrus clouds is no 'silver bullet' to avert global warming. There is the strong consensus of this being true for other measures as e. g. SAI, ocean acidification, or global dimming. However, a combination of several methods, when carefully applied on small scale and limited temporal scales, may possibly help to decelerate climate change, hence, neglecting severe side-effects, at least helps 'buying more time' for inalienably profound changes in everyday life considering emitting greenhouse gases and adjust to status quo.

Bibliography

Alterskjær, K., Kristjánsson, J. E., Boucher, O., Muri, H., Niemeier, U., Schmidt, H., Schulz, M., and Timmreck, C.: Sea - salt injections into the low - latitude marine boundary layer: The transient response in three Earth system models, J. Geophys. Res., 118, 12,195 – 12,206, 2013.

Andrews, R., Stein, J., Hansen, C., and Riley, D.: Introduction to the open source PV LIB for python Photovoltaic system modelling package, in: 2014 IEEE 40th Photovoltaic Specialist Conference, PVSC 2014, pp. 0170 – 0174, 2014.

Angel, R.: Feasibility of cooling the Earth with a cloud of small spacecraft near the inner Lagrange point (L1), Proc. Nat. Acad. Sci. USA, 103, 17 184 – 17 189, 2006.

Appleman, H.: The Formation of Exhaust Condensation Trails by Jet Aircraft, Bull. Amer. Meteor. Soc., 34, 14 – 20, 1953.

Atkinson, J. D., Murray, B. J., Woodhouse, M. T., Whale, T. F., Baustian, K. J., Carslaw, K. S., Dobbie, S., O'Sullivan, D., and Malkin, T. L.: The importance of feldspar for ice nucleation by mineral dust in mixed - phase clouds, Nature, 498, 339 – 340, 2013.

Baer, F., Arsky, N., Chamey, J. J., and Ellingson, R. G.: Intercomparison of heating rates generated by global climate model longwave radiation codes, J. Geophys. Res., 101, 26 589 – 26 603, 1996.

Baldauf, M., Seifert, A., Förstner, J., Majewski, D., Raschendorfer, M., and Reinhardt, T.: Operational Convective - Scale Numerical Weather Prediction with the COSMO Model: Description and Sensitivities, Mon. Wea. Rev., 139, 3887 – 3905, 2011.

Bangert, M., Nenes, A., Vogel, B., Vogel, H., Barahona, D., Karydis, V. A., Kumar, P., Kottmeier, C., and Blahak, U.: Saharan dust event impacts on cloud formation and radiation over Western Europe, Atmos. Chem. Phys., 12, 4045 – 4063, 2012.

Barahona, D. and Nenes, A.: Parameterizing the competition between homogeneous and heterogeneous freezing in cirrus cloud formation - monodisperse ice nuclei, Atmos. Chem. Phys., 9, 369 – 381, 2009a.

Barahona, D. and Nenes, A.: Parameterizing the competition between homogeneous and heterogeneous freezing in ice cloud formation - polydisperse ice nuclei, Atmos. Chem. Phys., 9, 5933 – 5948, 2009b.

Barahona, D. and Nenes, A.: Dynamical states of low temperature cirrus, Atmos. Chem. Phys., 11, 3757 – 3771, 2011.

Barahona, D., West, R. E. L., Stier, P., Romakkaniemi, S., Kokkola, H., and Nenes, A.: Comprehensively accounting for the effect of giant CCN in cloud activation parameterizations, Atmos. Chem. Phys., 10, 2467 – 2473, 2010.

Barahona, D., Molod, A., and Kalesse, H.: Direct estimation of the global distribution of vertical velocity within cirrus clouds, Sci. Rep., 7, 2017.

Binder, H., Boettcher, M., Grams, C. M., Joos, H., Pfahl, S., and Wernli, H.: Exceptional Air Mass Transport and Dynamical Drivers of an Extreme Wintertime Arctic Warm Event, Geophys. Res. Lett., 44, 12,028 – 12,036, 2017.

Bock, L. and Burkhardt, U.: The temporal evolution of a long - lived contrail cirrus cluster: Simulations with a global climate model, J. Geophys. Res., 121, 3548 – 3565, 2016a.

Bock, L. and Burkhardt, U.: Reassessing properties and radiative forcing of contrail cirrus using a climate model, J. Geophys. Res., 121, 9717 – 9736, 2016b.

Boose, Y., Welti, A., Atkinson, J., Ramelli, F., Danielczok, A., Bingemer, H. G., Plötze, M., Sierau, B., Kanji, Z. A., and Lohmann, U.: Heterogeneous ice nucleation on dust particles sourced from nine deserts worldwide - Part 1: Immersion freezing, Atmos. Chem. Phys., 16, 15 075 – 15 095, 2016.

Boucher, O., Randall, D., Artaxo, P., Bretherton, C., Feingold, G., Forster, P., Kerminen, V. M., Kondo, Y., Liao, H., Lohmann, U., Rasch, P., Satheesh, S., Sherwood, S., Stevens, B., and Zhang, X.: Climate Change 2013: The Physical Science Basis. Contribution of Working Group I to the Fifth Assessment Report of the Intergovernmental Panel on Climate Change, chap. Clouds and Aerosols, pp. 571 – 658, Cambridge University Press, Cambridge, United Kingdom and New York, NY, USA, 2013.

Boucher, O., Kleinschmitt, C., and Myhre, G.: Quasi - Additivity of the Radiative Effects of Marine Cloud Brightening and Stratospheric Sulfate Aerosol Injection, Geophys. Res. Lett., 44, 11,158 – 11,165, 2017.

Burkhardt, U. and Kärcher, B.: Process - based simulation of contrail cirrus in a global climate model, J. Geophys. Res., 114, 2009.

Burkhardt, U. and Kärcher, B.: Global radiative forcing from contrail cirrus, Nat. Clim. Change, 1, 54 – 58, 2011.

Caiazzo, F., Agarwal, A., Speth, R. L., and Barrett, S. R. H.: Impact of biofuels on contrail warming, Environ. Res. Lett., 12, 114 013, 2017.

Caldeira, K., Bala, G., and Cao, L.: The Science of Geoengineering, Annu. Rev. Earth Planet. Sci., 41, 231 – 256, 2013.

Chen, T., Rossow, W. B., and Zhang, Y.: Radiative Effects of Cloud - Type Variations, J. Climate, 13, 264 – 286, 2000.

Corti, T. and Peter, T.: A simple model for cloud radiative forcing, Atmos. Chem. Phys., 9, 5751 – 5758, 2009.

Crook, J. A., Jackson, L. S., Osprey, S. M., and Forster, P. M.: A comparison of temperature and precipitation responses to different Earth radiation management geoengineering schemes, J. Geophys. Res., 120, 9352 – 9373, 2015.

Crutzen, P. J.: Albedo Enhancement by Stratospheric Sulfur Injections: A Contribution to Resolve a Policy Dilemma?, Clim. Change, 77, 211, 2006.

Cziczo, D. J., Froyd, K. D., Hoose, C., Jensen, E. J., Diao, M., Zondlo, M. A., Smith, J. B., Twohy, C. H., and Murphy, D. M.: Clarifying the Dominant Sources and Mechanisms of Cirrus Cloud Formation, Science, 340, 1320 – 1324, 2013.

DeMott, P. J., Meyers, M. P., and Cotton, W. R.: Parameterization and Impact of Ice initiation Processes Relevant to Numerical Model Simulations of Cirrus Clouds, J. Geophys. Res., 51, 77 – 90, 1994.

DeMott, P. J., Cziczo, D. J., Prenni, A. J., Murphy, D. M., Kreidenweis, S. M., Thomson, D. S., Borys, R., and Rogers, D. C.: Measurements of the concentration and composition of nuclei for cirrus formation, Proc. Nat. Acad. Sci. USA, 100, 14 655 – 14 660, 2003.

Dessens, J.: A Physical Evaluation of a Hail Suppression Project with Silver Iodide Ground Burners in Southwestern France, Journal of Applied Meteorology, 37, 1588 – 1599, 1998.

Deuber, O., Matthes, S., Sausen, R., Ponater, M., and Lim, L.: A physical metric - based framework for evaluating the climate trade - off between CO2 and contrails - The case of lowering aircraft flight trajectories, Environmental Science & Policy, 25, 176 – 185, 2013.

Doms, G., Förstner, J., Heise, E., Herzog, H. J., Mironov, D., Raschendorfer, M., Reinhardt, T., Ritter, B., Schrodin, R., Schulz, J. P., and Vogel, G.: A Description of the Nonhydrostatic Regional COSMO - Model Part II: Physical Parameterization, Tech. Rep., Deutscher Wetterdienst, pp. 113 – 117, 2015a.

Doms, G., Schättler, U., and Baldauf, M.: A Description of the Nonhydrostatic Regional COSMO - Model Part I: Dynamics and Numerics., Tech. Rep., Deutscher Wetterdienst, 2015b.

Duda, D. P., Minnis, P., Nyuyen, L., and Palikonda, R.: A case study of the development of contrail clusters over the Great Lakes, J. Atmos. Sci., 61, 1132 – 1146, 2004.

Eastwood, M. L., Cremel, S., Gehrke, C., Girard, E., and Bertram, A. K.: Ice nucleation on mineral dust particles: Onset conditions, nucleation rates and contact angles, J. Geophys. Res., 113, 2008.

Eidhammer, T., Morrison, H., Bansemer, A., Gettelman, A., and Heymsfield, A. J.: Comparison of ice cloud properties simulated by the Community Atmosphere Model (CAM5) with in - situ observations, Atmos. Chem. Phys., 14, 10 103 – 10 118, 2014.

Eleftheratos, K., Zerefos, C. S., Zanis, P., Balis, D. S., Tselioudis, G., Gierens, K., and Sausen, R.: A study on natural and manmade global interannual fluctuations of cirrus cloud cover for the period 1984–2004, Atmos. Chem. Phys., 7, 2631 – 2642, 2007.

Fan, J., Ghan, S., Ovchinnikov, M., Liu, X., Rasch, P. J., and Korolev, A.: Representation of Arctic mixed - phase clouds and the Wegener - Bergeron - Findeisen process in climate models: Perspectives from a cloud - resolving study, J. Geophys. Res., 116, 2011.

Febvre, G., Gayet, J. F., Minikin, A., Schlager, H., Shcherbakov, V., Jourdan, O., Busen, R., Fiebig, M., Kärcher, B., and Schumann, U.: On optical and microphysical characteristics of contrails and cirrus, J. Geophys. Res., 114, 2009.

flightradar24.com: http://www.flightradar24.com, 16.08.2015, 2015.

Fountoukis, C. and Nenes, A.: Continued development of a cloud droplet formation parameterization for global climate models, J. Geophys. Res., 110, 2005.

Freudenthaler V., Homburg F., and Jäger H.: Contrail observations by ground - based scanning lidar: Cross - sectional growth, Geophys. Res. Lett., 22, 3501 – 3504, 1995.

Friedl Vallon, F., Gulde, T., Hase, F., Kleinert, A., Kulessa, T., Maucher, G., Neubert, T., Olschewski, F., Piesch, C., Preusse, P., Rongen, H., Sartorius, C., Schneider, H., Schönfeld, A., Tan, V., Bayer, N., Blank, J., Dapp, R., Ebersoldt, A., Fischer, H., Graf, F., Guggenmoser, T., Höpfner, M., Kaufmann, M., Kretschmer, E., Latzko, T., Nordmeyer, H., Oelhaf, H., Orphal, J., Riese, M., Schardt, G., Schillings, J., Sha, M. K., Suminska Ebersoldt, O., and Ungermann, J.: Instrument concept of the imaging Fourier transform spectrometer GLORIA, Atmos. Meas. Tech., 7, 3565 – 3577, 2014.

Fu, Q.: An Accurate Parameterization of the Solar Radiative Properties of Cirrus Clouds for Climate Models, J. Climate, 9, 2058 – 2082, 1996.

Fu, Q.: A New Parameterization of an Asymmetry Factor of Cirrus Clouds for Climate Models, J. Atmos. Sci., 64, 4140 – 4150, 2007.

Fu, Q., Yang, P., and Sun, W. B.: An Accurate Parameterization of the Infrared Radiative Properties of Cirrus Clouds for Climate Models, J. Climate, 11, 2223 – 2237, 1998.

Fusina, F., Spichtinger, P., and Lohmann, U.: Impact of ice supersaturated regions and thin cirrus on radiation in the midlatitudes, J. Geophys. Res., 112, 2007.

Futyan, J. M., Russell, J. E., and Harries, J. E.: Determining cloud forcing by cloud type from geostationary satellite data, Geophys. Res. Lett., 32, 2005.

Gasch, P., Rieger, D., Walter, C., Khain, P., Levi, Y., Knippertz, P., and Vogel, B.: Revealing the meteorological drivers of the September 2015 severe dust event in the Eastern Mediterranean, Atmos. Chem. Phys., 17, 13 573 – 13 604, 2017.

Gasparini, B. and Lohmann, U.: Why cirrus cloud seeding cannot substantially cool the planet, J. Geophys. Res., 121, 4877 – 4893, 2016.

Gasparini, B., Münch, S., Poncet, L., Feldmann, M., and Lohmann, U.: Is increasing ice crystal sedimentation velocity in geoengineering simulations a good proxy for cirrus cloud seeding?, Atmos. Chem. Phys., 17, 4871 – 4885, 2017.

Gaßmann, A.: 3D - transport of precipitation, COSMO Newsletter No. 2, Deutscher Wetterdienst, pp. 113 – 117, 2002.

Gaßmann, A.: Case Studies with the 2 - Timelevel Scheme and Prognostic Precipitation, COSMO Newsletter No. 3, Deutscher Wetterdienst, pp. 173 – 176, 2003.

Gettelman, A., Liu, X., Barahona, D., Lohmann, U., and Chen, C.: Climate impacts of ice nucleation, J. Geophys. Res., 117, 2012.

Gruber, S.: Simulating Contrails and Their Impact on Incoming Solar Radiation at the Surface on the Regional Scale - A Case Study, Master's thesis, Institute of Meteorology and Climate Research, Karlsruhe Institute of Technology (KIT), 2015.

Gruber, S., Unterstrasser, S., Bechtold, J., Vogel, H., Jung, M., Pak, H., and Vogel, B.: Contrails and their impact on shortwave radiation and photovoltaic power production - a regional model study, Atmos. Chem. Phys., 18, 6393 – 6411, 2018.

Gupta, S. K., Whitlock, C. H., Ritchey, N. A., and Wilber, A. C.: Cloud and the Earth's Radiant Energy System (CERES) Algorithm Theoretical Basis Document, chap. An Algorithm for Longwave Surface Radiation Budget for Total Skies, CERES ATBD Subsystem 4.6.3 Release 2.2, 1997.

Haag, W. and Kärcher, B.: The impact of aerosols and gravity waves on cirrus clouds at midlatitudes, J. Geophys. Res., 109, 2004.

Harrington, J. Y., Meyers, M. P., Walko, R. L., and Cotton, W. R.: Parameterization of Ice Crystal Conversion Processes Due to Vapor Deposition for Mesoscale Models Using Double - Moment Basis Functions. Part I: Basic Formulation and Parcel Model Results, J. Atmos. Sci., 52, 4344 – 4366, 1995.

Harrington, J. Y., Reisin, T., Cotton, W. R., and Kreidenweis, S. M.: Cloud resolving simulations of Arctic stratus: Part II: Transition - season clouds, Atmos. Res., 51, 45 – 75, 1999.

Hartmann, D. L., Bell, M. E. O., and Michelsen, M. L.: The Effect of Cloud Type on Earth's Energy Balance: Global Analysis, J. Climate, 5, 1281 – 1304, 1992.

Heinze, R., Dipankar, A., Henken, C. C., Moseley, C., Sourdeval, O., Trömel, S., Xie, X., Adamidis, P., Ament, F., Baars, H., Barthlott, C., Behrendt,

A., Blahak, U., Bley, S., Brdar, S., Brueck, M., Crewell, S., Deneke, H., Girolamo, P. D., Evaristo, R., Fischer, J., Frank, C., Friederichs, P., Göcke, T., Gorges, K., Hande, L., Hanke, M., Hansen, A., Hege, H.-C., Hoose, C., Jahns, T., Kalthoff, N., Klocke, D., Kneifel, S., Knippertz, P., Kuhn, A., van Laar, T., Macke, A., Maurer, V., Mayer, B., Meyer, C. I., Muppa, S. K., Neggers, R. A. J., Orlandi, E., Pantillon, F., Pospichal, B., Röber, N., Scheck, L., Seifert, A., Seifert, P., Senf, F., Siligam, P., Simmer, C., Steinke, S., Stevens, B., Wapler, K., Weniger, M., Wulfmeyer, V., Zängl, G., Zhang, D., and Quaas, J.: Large - eddy simulations over Germany using ICON: a comprehensive evaluation, Quart. J. Roy. Meteorol. Soc., 143, 69 – 100, 2017.

Heymsfield, A., Baumgardner, D., DeMott, P., Forster, P., Gierens, K., and Kärcher, B.: Contrail Microphysics, Bull. Amer. Meteor. Soc., 91, 465 – 472, 2010.

Heymsfield, A. J. and Iaquinta, J.: Cirrus Crystal Terminal Velocities, J. Atmos. Sci., 57, 916 – 938, 2000.

Heymsfield, A. J., Lawson, R. P., and Sachse, G. W.: Growth of ice crystals in a precipitating contrail, Geophys. Res. Lett., 25, 114 013, 1998.

Heymsfield, A. J., Field, P., and Bansemer, A.: Exponential Size Distributions for Snow, J. Atmos. Sci., 65, 4017 – 4031, 2008.

Holland, M. M. and Bitz, C. M.: Polar amplification of climate change in coupled models, Climate Dynamics, 21, 221 – 232, 2003.

Hong, Y., Liu, G., and Li, J. L. F.: Assessing the Radiative Effects of Global Ice Clouds Based on CloudSat and CALIPSO Measurements, J. Climate, 29, 7651 – 7674, 2016.

Hoose, C. and Möhler, O.: Heterogeneous ice nucleation on atmospheric aerosols: a review of results from laboratory experiments, Atmos. Chem. Phys., 12, 9817 – 9854, 2012.

Hu, Y. X. and Stamnes, K.: An Accurate Parameterization of the Radiative Properties of Water Clouds Suitable for Use in Climate Models, J. Climate, 6, 728 – 742, 1993.

Hunt, W. H., Winker, D. M., Vaughan, M. A., Powell, K. A., Lucker, P. L., and Weimer, C.: CALIPSO Lidar Description and Performance Assessment, J. Atmos. Ocean. Technol., 26, 1214 – 1228, 2009.

Ickes, L., Welti, A., Hoose, C., and Lohmann, U.: Classical nucleation theory of homogeneous freezing of water: thermodynamic and kinetic parameters, Phys. Chem. Chem. Phys., 17, 5514 – 5537, 2015.

Inman, R. H., Pedro, H. T., and Coimbra, C. F.: Solar forecasting methods for renewable energy integration, Prog. Energy Combust. Sci., 39, 535 – 576, 2013.

Irvine, E. A. and Shine, K. P.: Ice supersaturation and the potential for contrail formation in a changing climate, Earth System Dynamics, 6, 555 – 568, 2015.

Irvine, P. J., Ridgwell, A., and Lunt, D. J.: Climatic effects of surface albedo geoengineering, J. Geophys. Res., 116, 2011.

Irvine, P. J., Kravitz, B., Lawrence, M. G., and Muri, H.: An overview of the Earth system science of solar geoengineering, Wiley Interdisciplinary Reviews: Climate Change, 7, 815 – 833, 2016.

Iwabuchi, H., Yang, P., Liou, K. N., and Minnis, P.: Physical and optical properties of persistent contrails: Climatology and interpretation, J. Geophys. Res., 102, 114 013, 2012.

Jackson, L. S., Crook, J. A., and Forster, P. M.: An intensified hydrological cycle in the simulation of geoengineering by cirrus cloud thinning using ice crystal fall speed changes, J. Geophys. Res., 121, 6822 – 6840, 2016.

Jensen, E. and Pfister, L.: Transport and freeze - drying in the tropical tropopause layer, J. Geophys. Res., 109, 2004.

Jensen, E. J., Pfister, L., Bui, T. P., Lawson, P., and Baumgardner, D.: Ice nucleation and cloud microphysical properties in tropical tropopause layer cirrus, Atmos. Chem. Phys., 10, 1369 – 1384, 2010.

Jensen, E. J., Diskin, G., Lawson, R. P., Lance, S., Bui, T. P., Hlavka, D., McGill, M., Pfister, L., Toon, O. B., and Gao, R.: Ice nucleation and dehydration in the Tropical Tropopause Layer, Proc. Nat. Acad. Sci. USA, 110, 2041 – 2046, 2013.

Jensen, E. J., Ueyama, R., Pfister, L., Bui, T. V., Alexander, M. J., Podglajen, A., Hertzog, A., Woods, S., Lawson, R. P., Kim, J. E., and Schoeberl, M. R.: High - frequency gravity waves and homogeneous ice nucleation in tropical tropopause layer cirrus, Geophys. Res. Lett., 43, 6629 – 6635, 2016.

Johansson, S., Woiwode, W., Höpfner, M., Friedl Vallon, F., Kleinert, A., Kretschmer, E., Latzko, T., Orphal, J., Preusse, P., Ungermann, J., Santee, M. L., Jurkat Witschas, T., Marsing, A., Voigt, C., Giez, A., Krämer, M., Rolf, C., Zahn, A., Engel, A., Sinnhuber, B. M., and Oelhaf, H.: Airborne limb - imaging measurements of temperature, HNO_3, O_3, $ClONO_2$, H_2O and CFC - 12 during the Arctic winter 2015/16: characterization, in - situ validation and comparison to Aura/MLS, Atmospheric Measurement Techniques Discussions, 2018, 1 – 31, 2018.

Jones, A., Haywood, J. M., Alterskjær, K., Boucher, O., Cole, J. N. S., Curry, C. L., Irvine, P. J., Ji, D., Kravitz, B., Egill Kristjánsson, J., Moore, J. C., Niemeier, U., Robock, A., Schmidt, H., Singh, B., Tilmes, S., Watanabe, S., and Yoon, J. H.: The impact of abrupt suspension of solar radiation management (termination effect) in experiment G2 of the Geoengineering Model Intercomparison Project (GeoMIP), J. Geophys. Res., 118, 9743 – 9752, 2013.

Joos, H., Spichtinger, P., Reutter, P., and Fusina, F.: Influence of heterogeneous freezing on the microphysical and radiative properties of orographic cirrus clouds, Atmos. Chem. Phys., 14, 6835 – 6852, 2014.

Kalesse, H. and Kollias, P.: Climatology of High Cloud Dynamics Using Profiling ARM Doppler Radar Observations, J. Climate, 26, 6340 – 6359, 2013.

Kärcher, B. and Lohmann, U.: A parameterization of cirrus cloud formation: Homogeneous freezing of supercooled aerosols, J. Geophys. Res., 107, AAC 4 – 1 – AAC 4 – 10, 2002.

Kärcher, B. and Lohmann, U.: A parameterization of cirrus cloud formation: Heterogeneous freezing, J. Geophys. Res., 108, 2003.

Kärcher, B. and Ström, J.: The roles of dynamical variability and aerosols in cirrus cloud formation, Atmos. Chem. Phys., 3, 823 – 838, 2003.

Kärcher, B., Hendricks, J., and Lohmann, U.: Physically based parameterization of cirrus cloud formation for use in global atmospheric models, J. Geophys. Res., 111, 2006.

Kärcher, B., Burkhardt, U., Ponater, M., and Frömming, C.: Importance of representing optical depth variability for estimates of global line - shaped contrail radiative forcing, Proc. Nat. Acad. Sci. USA, 107, 19 181 – 19 184, 2010.

Kärcher, B., Burkhardt, U., Bier, A., Bock, L., and Ford, I. J.: The microphysical pathway to contrail formation, J. Geophys. Res., 120, 7893 – 7927, 2015.

Key, J. R., Yang, P., Baum, B. A., and Nasiri, S. L.: Parameterization of shortwave ice cloud optical properties for various particle habits, J. Geophys. Res., 107, AAC 7 – 1 – AAC 7 – 10, 2002.

Khosrawi, F., Kirner, O., Sinnhuber, B. M., Johansson, S., Höpfner, M., Santee, M. L., Froidevaux, L., Ungermann, J., Ruhnke, R., Woiwode, W., Oelhaf, H., and Braesicke, P.: Denitrification, dehydration and ozone loss during the 2015/2016 Arctic winter, Atmos. Chem. Phys., 17, 12 893 – 12 910, 2017.

Khou, J. C., Ghedhaïfi, W., Vancassel, X., and Garnier, F.: Spatial Simulation of Contrail Formation in Near - Field of Commercial Aircraft, J. Aircraft, 52, 1927 – 1938, 2015.

Kienast Sjögren, E., Rolf, C., Seifert, P., Krieger, U. K., Luo, B. P., Krämer, M., and Peter, T.: Climatological and radiative properties of midlatitude cirrus clouds derived by automatic evaluation of lidar measurements, Atmos. Chem. Phys., 16, 7605 – 7621, 2016.

Kiselev, A., Bachmann, F., Pedevilla, P., Cox, S. J., Michaelides, A., Gerthsen, D., and Leisner, T.: Active sites in heterogeneous ice nucleation - the example of K - rich feldspars, Science, 355, 367 – 371, 2017.

Kleinschmitt, C., Boucher, O., and Platt, U.: Sensitivity of the radiative forcing by stratospheric sulfur geoengineering to the amount and strategy of the SO2 injection studied with the LMDZ - S3A model, Atmos. Chem. Phys., 18, 2769 – 2786, 2018.

Köhler, C., Steiner, A., Drenan, Y. M. S., Ernst, D., Dick, A. B., Zirkelbach, M., Bouallègue, Z. B., Metzinger, I., and Ritter, B.: Critical weather situations for renewable energies - Part B: Low stratus risk for solar power, Renew. Energ., 101, 535 – 576, 2017.

Köhler, C. G. and Seifert, A.: Identifying sensitivities for cirrus modelling using a two - moment two - mode bulk microphysics scheme, Tellus, 67, 24 494, 2015.

Koop, T., Luo, B., Tsias, A., and Peter, T.: Water activity as the determinant for homogeneous ice nucleation in aqueous solutions, Nature, 406, 611 – 614, 2000.

Korolev, A. V., Emery, E. F., Strapp, J. W., Cober, S. G., Isaac, G. A., Wasey, M., and Marcotte, D.: Small Ice Particles in Tropospheric Clouds: Fact or Artifact? Airborne Icing Instrumentation Evaluation Experiment, Bull. Amer. Meteor. Soc., 92, 967 – 973, 2011.

Krämer, M., Schiller, C., Afchine, A., Bauer, R., Gensch, I., Mangold, A., Schlicht, S., Spelten, N., Sitnikov, N., Borrmann, S., de Reus, M., and Spichtinger, P.: Ice supersaturations and cirrus cloud crystal numbers, Atmos. Chem. Phys., 9, 3505 – 3522, 2009.

Krämer, M., Rolf, C., Luebke, A., Afchine, A., Spelten, N., Costa, A., Meyer, J., Zöger, M., Smith, J., Herman, R. L., Buchholz, B., Ebert, V., Baumgardner, D., Borrmann, S., Klingebiel, M., and Avallone, L.: A microphysics guide to cirrus clouds - Part 1: Cirrus types, Atmos. Chem. Phys., 16, 3463 – 3483, 2016.

Kravitz, B., Caldeira, K., Boucher, O., Robock, A., Rasch, P. J., Alterskjær, K., Karam, D. B., Cole, J. N. S., Curry, C. L., Haywood, J. M., Irvine, P. J., Ji, D., Jones, A., Kristjánsson, J. E., Lunt, D. J., Moore, J. C., Niemeier, U., Schmidt, H., Schulz, M., Singh, B., Tilmes, S., Watanabe, S., Yang, S., and Yoon, J.-H.: Climate model response from the Geoengineering Model Intercomparison Project (GeoMIP), J. Geophys. Res., 118, 8320 – 8332, 2013.

Kristjánsson, J. E., Muri, H., and Schmidt, H.: The hydrological cycle response to cirrus cloud thinning, Geophys. Res. Lett., 42, 10,807 – 10,815, 2015.

Kuebbeler, M., Lohmann, U., Hendricks, J., and Kärcher, B.: Dust ice nuclei effects on cirrus clouds, Atmos. Chem. Phys., 14, 3027 – 3046, 2014.

Kumar, P., Sokolik, I. N., and Nenes, A.: Parameterization of cloud droplet formation for global and regional models: including adsorption activation from insoluble CCN, Atmos. Chem. Phys., 9, 2517 – 2532, 2009.

Lamquin, N., Stubenrauch, C. J., Gierens, K., Burkhardt, U., and Smit, H.: A global climatology of upper - tropospheric ice supersaturation occurrence inferred from the Atmospheric Infrared Sounder calibrated by MOZAIC, Atmos. Chem. Phys., 12, 381 – 405, 2012.

Latham, J.: Control of global warming?, Nature, 347, 339 – 340, 1990.

Lew, D. and Richard, P.: Western wind and solar integration study, Tech. rep., National Renewable Energy Laboratories, 2010.

Lewellen, D. C.: Persistent Contrails and Contrail Cirrus. Part II: Full Lifetime Behavior, J. Atmos. Sci., 71, 4420 – 4438, 2014.

Lewellen, D. C. and Lewellen, W. S.: The Effects of Aircraft Wake Dynamics on Contrail Development, J. Atmos. Sci., 58, 390 – 406, 2001.

Lewellen, D. C., Meza, O., and Huebsch, W. W.: Persistent Contrails and Contrail Cirrus. Part I: Large - Eddy Simulations from Inception to Demise, J. Atmos. Sci., 71, 4399 – 4419, 2014.

Loeb, N. G., Wielicki, B. A., Doelling, D. R., Smith, G. L., Keyes, D. F., Kato, S., Manalo Smith, N., and Wong, T.: Toward Optimal Closure of the Earth's Top - of - Atmosphere Radiation Budget, Journal of Climate, 22, 748 – 766, 2009.

Lohmann, U. and Gasparini, B.: A cirrus cloud climate dial?, Science, 357, 248 – 249, 2017.

Lohmann, U., McFarlane, N., Levkov, L., Abdella, K., and Albers, F.: Comparing Different Cloud Schemes of a Single Column Model by Using Mesoscale Forcing and Nudging Technique, J. Climate, 12, 438 – 461, 1999.

Lohmann, U., Spichtinger, P., Jess, S., Peter, T., and Smit, H.: Cirrus cloud formation and ice supersaturated regions in a global climate model, Environ. Res., 3, 045 022, 2008.

Luebke, A. E., Afchine, A., Costa, A., Grooß, J. U., Meyer, J., Rolf, C., Spelten, N., Avallone, L. M., Baumgardner, D., and Krämer, M.: The origin of midlatitude ice clouds and the resulting influence on their microphysical properties, Atmos. Chem. Phys., 16, 5793 – 5809, 2016.

Lundgren, K., Vogel, B., Vogel, H., and Kottmeier, C.: Direct radiative effects of sea salt for the Mediterranean region under conditions of low to moderate wind speeds, J. Geophys. Res., 118, 1906 – 1923, 2013.

Mannstein, H., Spichtinger, P., and Gierens, K.: A note on how to avoid contrail cirrus, Transportation Research Part D: Transport and Environment, 10, 421 – 426, 2005.

Marquart, S., Ponater, M., Mager, F., and Sausen, R.: Future Development of Contrail Cover, Optical Depth, and Radiative Forcing: Impacts of Increasing Air Traffic and Climate Change, J. Climate, 16, 2890 – 2904, 2003.

Matus, A. V. and L'Ecuyer, T. S.: The role of cloud phase in Earth's radiation budget, J. Geophys. Res., 122, 2559 – 2578, 2017.

McFarlane, S. A., Mather, J. H., and Ackerman, T. P.: Analysis of tropical radiative heating profiles: A comparison of models and observations, J. Geophys. Res., 112, 2007.

Minnis, P., Young, D. F., Garber, D. P., Nguyen, L., Smith, W. L., and Palikonda, R.: Transformation of contrails into cirrus during SUCCESS, Geophys. Res. Lett., 25, 1157 – 1160, 1998.

Mitchell, D., Mishra, S., and Paul Lawson, R.: Planet Earth 2011 - Global Warming Challenges and Opportunities for Policy and Practice, vol. 12, chap. Cirrus Clouds and Climate Engineering: New Findings on Ice Nucleation and Theoretical Basis, pp. 257 – 288, InTech., Rijeka, Croatia, 2011.

Mitchell, D. L. and Finnegan, W.: Modification of cirrus clouds to reduce global warming, Environ. Res., 4, 045 102, 2009.

Mlawer, E. J., Taubman, S. J., Brown, P. D., Iacono, M. J., and Clough, S. A.: Radiative transfer for inhomogeneous atmospheres: RRTM, a validated correlated - k model for the longwave, J. Geophys. Res., 102, 16 663 – 16 682, 1997.

Moore, R. H., Thornhill, K. L., Weinzierl, B., Sauer, D., D'Ascoli, E., Kim, J., Lichtenstern, M., Scheibe, M., Beaton, B., Beyersdorf, A. J., Barrick, J., Bulzan, D., Corr, C. A., Crosbie, E., Jurkat, T., Martin, R., Riddick, D., Shook, M., Slover, G., Voigt, C., White, R., Winstead, E., Yasky, R., Ziemba, L. D., Brown, A., Schlager, H., and Anderson, B. E.: Biofuel blending reduces particle emissions from aircraft engines at cruise conditions, Nature, 543, 411 – 415, 2017.

Morales, R. and Nenes, A.: Characteristic updrafts for computing distribution - averaged cloud droplet number and stratocumulus cloud properties, J. Geophys. Res., 115, 2010.

Muri, H., Kristjánsson, J. E., Storelvmo, T., and Pfeffer, M. A.: The climatic effects of modifying cirrus clouds in a climate engineering framework, J. Geophys. Res., 119, 4174 – 4191, 2014.

Myhre, G., Shindell, D., Bréon, F. M., Collins, W., Fuglestvedt, J., Huang, J., Koch, D., Lamarque, J. F., Lee, D., Mendoza, B., Nakajima, T., Robock, A., Stephens, G., Takemura, T., and Zhan, H.: Climate Change 2013: The Physical Science Basis. Contribution of Working Group I to the Fifth Assessment Report of the Intergovernmental Panel on Climate Change, chap. Anthropogenic and Natural Radiative Forcing, pp. 659 – 740, Cambridge University Press, Cambridge, United Kingdom and New York, NY, USA, 2013.

NASA/GSFC/ESDIS: 2018.

Niemand, M., Möhler, O., Vogel, B., Vogel, H., Hoose, C., Connolly, P., Klein, H., Bingemer, H., DeMott, P., Skrotzki, J., and Leisner, T.: A Particle - Surface - Area - Based Parameterization of Immersion Freezing on Desert Dust Particles, J. Atmos. Sci., 69, 3077 – 3092, 2012.

Paoli, R., Nybelen, L., Picot, J., and Cariolle, D.: Effects of jet/vortex interaction on contrail formation in supersaturated conditions, Phys. Fluids, 25, 1 – 28, 2013.

Parker, A. and Irvine, P. J.: The Risk of Termination Shock From Solar Geoengineering, Earth's Future, 6, 456 – 467, 2018.

Penner, J. E., Zhou, C., and Liu, X.: Can cirrus cloud seeding be used for geoengineering?, Geophys. Res. Lett., 42, 8775 – 8782, 2015.

Petty, G.: A First Course in Atmospheric Radiation, Sundog Publishing, Madison, 2004.

Petzold, A., Busen, R., Schröder, F. P., Baumann, R., Kuhn, M., Ström, J., Hagen, D. E., Whitefield, P. D., Baumgardner, D., Arnold, F., Borrmann, S., and Schumann, U.: Near - field measurements on contrail properties from fuels with different sulfur content, J. Geophys. Res., 102, 114 013, 1997.

Phillips, V. T. J., DeMott, P. J., and Andronache, C.: An Empirical Parameterization of Heterogeneous Ice Nucleation for Multiple Chemical Species of Aerosol, J. Atmos. Sci., 65, 2757 – 2783, 2008.

Phillips, V. T. J., Demott, P. J., Andronache, C., Pratt, K. A., Prather, K. A., Subramanian, R., and Twohy, C.: Improvements to an Empirical Parameterization of Heterogeneous Ice Nucleation and Its Comparison with Observations, J. Atmos. Sci., 70, 378 – 409, 2013.

Poellot, M. R., Arnott, W. P., and Hallett, J.: In situ observations of contrail microphysics and implications for their radiative impact, J. Geophys. Res., 104, 12 077 – 12 084, 1999.

Powell, K. A., Hostetler, C. A., Vaughan, M. A., Lee, K. P., Trepte, C. R., Rogers, R. R., Winker, D. M., Liu, Z., Kuehn, R. E., Hunt, W. H., and Young, S. A.: CALIPSO Lidar Calibration Algorithms. Part I: Nighttime 532 - nm Parallel Channel and 532 - nm Perpendicular Channel, J. Atmos. Ocean. Technol., 26, 2015 – 2033, 2009.

Pruppacher, H. R. and Klett, J. D.: Microphysics of clouds and precipitation, Dordrecht: Kluwer Academic Publishers, 1997.

Quante, M.: Turbulenz in Cirruswolken mittlerer Breiten, Ph.D. thesis, Hamburg, Univ., FB Geowiss., 2006.

Rauscher, S. A., O'Brien, T. A., Piani, C., Coppola, E., Giorgi, F., Collins, W. D., and Lawston, P. M.: A multimodel intercomparison of resolution effects on precipitation: simulations and theory, Clim. Dynam., 47, 2205 – 2218, 2016.

Ridgwell, A., Singarayer, J. S., Hetherington, A. M., and Valdes, P. J.: Tackling Regional Climate Change By Leaf Albedo Bio - geoengineering, Current Biology, 19, 146 – 150, 2009.

Rieger, D., Bangert, M., Bischoff Gauss, I., Förstner, J., Lundgren, K., Reinert, D., Schröter, J., Vogel, H., Zängl, G., Ruhnke, R., and Vogel, B.: ICON - ART 1.0 - a new online - coupled model system from the global to regional scale, Geosci. Model Dev., 8, 1659 – 1676, 2015.

Rieger, D., Steiner, A., Bachmann, V., Gasch, P., Förstner, J., Deetz, K., Vogel, B., and Vogel, H.: Impact of the 4 April 2014 Saharan dust outbreak on the photovoltaic power generation in Germany, Atmos. Chem. Phys., 17, 13 391 – 13 415, 2017.

Riese, M., Oelhaf, H., Preusse, P., Blank, J., Ern, M., Friedl Vallon, F., Fischer, H., Guggenmoser, T., Höpfner, M., Hoor, P., Kaufmann, M., Orphal, J., Plöger, F., Spang, R., Suminska Ebersoldt, O., Ungermann, J., Vogel, B., and Woiwode, W.: Gimballed Limb Observer for Radiance Imaging of the Atmosphere (GLORIA) scientific objectives, Atmos. Meas. Tech., 7, 1915 – 1928, 2014.

Ritter, B. and Geleyn, J. F.: A Comprehensive Radiation Scheme for Numerical Weather Prediction Models with Potential Applications in Climate Simulations, Mon. Wea. Rev., 120, 303 – 325, 1992.

Rossow, W. B. and Schiffer, R. A.: Advances in Understanding Clouds from ISCCP, Bull. Amer. Meteor. Soc., 80, 2261 – 2288, 1999.

S., U. and N., G.: Aircraft - type dependency of contrail evolution, J. Geophys. Res., 119, 14,015 – 14,027, 2014.

Sanderson, B. M., O'Neill, B. C., and Tebaldi, C.: What would it take to achieve the Paris temperature targets?, Geophys. Res. Lett., 43, 7133 – 7142, 2016.

Sassen, K., Wang, Z., and Liu, D.: Global distribution of cirrus clouds from CloudSat/Cloud - Aerosol Lidar and Infrared Pathfinder Satellite Observations (CALIPSO) measurements, J. Geophys. Res., 113, 2008.

Sausen, R., Isaksen, I., Grewe, V., Hauglustaine, D., Lee, D. S., Myhre, G., Köhler, M. O., Pitari, G., Schumann, U., Stordal, F., and Zerefos, C.: Aviation radiative forcing in 2000: An update on IPCC (1999), Meteor. Z., 14, 555 – 561, 2005.

Schmidt, C. T. and Garrett, T. J.: A Simple Framework for the Dynamic Response of Cirrus Clouds to Local Diabatic Radiative Heating, J. Atmos. Sci., 70, 1409 – 1422, 2013.

Schmidt, E.: Die Entstehung von Eisnebel aus den Auspuffgasen von Flugmotoren, Schriften der Deutschen Akademie der Luftfahrtforschung, 44, 1 – 15, 1941.

Schröder, F., Kärcher, B., Duroure, C., Ström, J., Petzold, A., Gayet, J. F., Strauss, B., Wendling, P., and Borrmann, S.: On the Transition of Contrails into Cirrus Clouds, J. Atmos. Sci., 57, 464 – 480, 2000.

Schumann, U.: On conditions for contrail formation from aircraft exhausts, Meteor. Z., 5, 4 – 23, 1996.

Schumann, U.: A contrail cirrus prediction model, Geosci. Model Dev., 5, 543 – 580, 2012.

Schumann, U. and Graf, K.: Aviation - induced cirrus and radiation changes at diurnal timescales, J. Geophys. Res., 118, 2404 – 2421, 2013.

Schumann, U. and Heymsfield, A.: On the lifecycle of individual contrails and contrail cirrus, Meteor. Monogr., 58, 3.1 – 3.24, 2017.

Schumann, U., Penner, J. E., Chen, Y., Zhou, C., and Graf, K.: Dehydration effects from contrails in a coupled contrail - climate model, Atmos. Chem. Phys., 15, 11 179 – 11 199, 2015.

Schumann, U., Baumann, R., amd S. T. Bedka, D. B., Duda, D. P., Freudenthaler, V., Gayet, J. F., Heymsfield, A. J., Minnis, P., Quante, M., Raschke, E., Schlager, H., Navarro, M. V., Voigt, C., and Wang, Z.: Properties of individual contrails: A compilation of observations and some comparisons, Atmos. Chem. Phys., 17, 403 – 438, 2017.

Seifert, A. and Beheng, K. D.: A two - moment cloud microphysics parameterization for mixed - phase clouds. Part 1: Model description, Meteorol. Atmos. Phys., 92, 45 – 66, 2006.

Shepherd, J.: Geoengineering the climate: Science, governance and uncertainty, Project report, The Royal Society, 2009.

Sheyko, B. A., Sullivan, S. C., Morales, R., Capps, S. L., Barahona, D., Shi, X., Liu, X., and Nenes, A.: Quantifying sensitivities of ice crystal number and sources of ice crystal number variability in CAM 5.1 using the adjoint of a physically based cirrus formation parameterization, J. Geophys. Res., 120, 2834 – 2854, 2015.

Shi, X. and Liu, X.: Effect of cloud - scale vertical velocity on the contribution of homogeneous nucleation to cirrus formation and radiative forcing, Geophys. Res. Lett., 43, 6588 – 6595, 2016.

Soler, M., Zou, B., and Hansen, M.: Flight trajectory design in the presence of contrails: Application of a multiphase mixed - integer optimal control approach, Transportation Research Part C: Emerging Technologies, 48, 172 – 194, 2014.

Spang, R., Remedios, J., and Barkley, M.: Colour indices for the detection and differentiation of cloud types in infra - red limb emission spectra, Advances in Space Research, 33, 1041 – 1047, 2004.

Spangenberg, D. A., Minnis, P., Bedka, S. T., Palikonda, R., Duda, D. P., and Rose, F. G.: Contrail radiative forcing over the Northern Hemisphere from 2006 Aqua MODIS data, Geophys. Res. Lett., 40, 595 – 600, 2013.

Spichtinger, P. and Gierens, K. M.: Modelling of cirrus clouds - Part 1a: Model description and validation, Atmos. Chem. Phys., 9, 685 – 706, 2009.

Spichtinger, P. and Krämer, M.: Tropical tropopause ice clouds: a dynamic approach to the mystery of low crystal numbers, Atmos. Chem. Phys., 13, 9801 – 9818, 2013.

Spichtinger, P., Gierens, K., and Read, W.: The global distribution of ice - supersaturated regions as seen by the Microwave Limb Sounder, Quart. J. Roy. Meteorol. Soc., 129, 3391 – 3410, 2003.

Spichtinger, P., Gierens, K., Smit, H. G. J., Ovarlez, J., and Gayet, J. F.: On the distribution of relative humidity in cirrus clouds, Atmos. Chem. Phys., 4, 639 – 647, 2004.

Stephens, G. L., Vane, D. G., Boain, R. J., Mace, G. G., Sassen, K., Wang, Z., Illingworth, A. J., O'connor, E. J., Rossow, W. B., Durden, S. L., Miller, S. D., Austin, R. T., Benedetti, A., Mitrescu, C., and the CloudSat Science Team: The CloudSat mission and the A - Train: A new dimension of space - based observations of clouds and precipitation, Bull. Amer. Meteor. Soc., 83, 1771 – 1790, 2002.

Stevens, B., Giorgetta, M., Esch, M., Mauritsen, T., Crueger, T., Rast, S., Salzmann, M., Schmidt, H., Bader, J., Block, K., Brokopf, R., Fast, I., Kinne, S., Kornblueh, L., Lohmann, U., Pincus, R., Reichler, T., and Roeckner, E.: Atmospheric component of the MPI - M Earth System Model: ECHAM6, J. Adv. Model. Earth Syst., 5, 146 – 172, 2013.

Stocker, T., Qin, D., Plattner, G. K., Alexander, L., Allen, S., Bindoff, N., Bréon, F. M., Church, J., Cubasch, U., Emori, S., Forster, P., Friedlingstein, P., Gillett, N., Gregory, J., Hartmann, D., Jansen, E., Kirtman, B., Knutti, R., Kumar, K. K., Lemke, P., Marotzke, J., Delmotte, V. M., Meehl, G., Mokhov, I., Piao, S., Ramaswamy, V., Randall, D., Rhein, M., Rojas, M., Sabine, C., Shindell, D., Talley, L., Vaughan, D., and Xie, S. P.: Climate Change 2013: The Physical Science Basis. Contribution of Working Group I to the Fifth Assessment Report of the Intergovernmental Panel on Climate Change, chap. Technical Summary, pp. 33 – 115, Cambridge University Press, Cambridge, United Kingdom and New York, NY, USA, 2013.

Storelvmo, T. and Herger, N.: Cirrus cloud susceptibility to the injection of ice nuclei in the upper troposphere, J. Geophys. Res., 119, 2375 – 2389, 2014.

Storelvmo, T., Kristjansson, J. E., Muri, H., Pfeffer, M., Barahona, D., and Nenes, A.: Cirrus cloud seeding has potential to cool climate, Geophys. Res. Lett., 40, 178 – 182, 2013.

Storelvmo, T., Boos, W. R., and Herger, N.: Cirrus cloud seeding: a climate engineering mechanism with reduced side effects?, Philos. Trans. Roy. Soc. London, 372, 2014.

Stubenrauch, C. J. and Schumann, U.: Impact of air traffic on cirrus coverage, Geophys. Res. Lett., 32, 2005.

Stuber, N. and Forster, P.: The impact of diurnal variations of air traffic on contrail radiative forcing, Atmos. Chem. Phys., 7, 3153 – 3162, 2007.

Unterstrasser, S.: Large - eddy simulation study of contrail microphysics and geometry during the vortex phase and consequences on contrail - to - cirrus transition, J. Geophys. Res., 119, 7537 – 7555, 2014.

Unterstrasser, S.: Properties of young contrails - a parametrisation based on large - eddy simulations, Atmos. Chem. Phys., 16, 2059 – 2082, 2016.

Unterstrasser, S. and Gierens, K.: Numerical simulations of contrail - to - cirrus transition - Part 1: An extensive parametric study, Atmos. Chem. Phys., 10, 2017 – 2036, 2010a.

Unterstrasser, S. and Gierens, K.: Numerical simulations of contrail - to - cirrus transition - Part 2: Impact of initial ice crystal number, radiation, stratification, secondary nucleation and layer depth, Atmos. Chem. Phys., 10, 2037 – 2051, 2010b.

Unterstrasser, S., Gierens, K., Sölch, I., and Lainer, M.: Numerical simulations of homogeneously nucleated natural cirrus and contrail - cirrus. Part 1: How different are they?, Meteor. Z., 2017a.

Unterstrasser, S., Gierens, K., Sölch, I., and Wirth, M.: Numerical simulations of homogeneously nucleated natural cirrus and contrail - cirrus. Part 2: Interaction on local scale, Meteor. Z., 2017b.

UWYO: http://weather.uwyo.edu/upperair/sounding.html, 01.02.2018, 2018.

Vázquez Navarro, M., Mannstein, H., and Kox, S.: Contrail life cycle and properties from 1 year of MSG/SEVIRI rapid - scan images, Atmos. Chem. Phys., 15, 8739 – 8749, 2015.

Vogel, B., Vogel, H., Bäumer, D., Bangert, M., Lundgren, K., Rinke, R., and Stanelle, T.: The comprehensive model system COSMO - ART - Radiative impact of aerosol on the state of the atmosphere on the regional scale, Atmos. Chem. Phys., 9, 8661 – 8680, 2009.

Voigt, C., Schumann, U., Jurkat, T., Schäuble, D., Schlager, H., Petzold, A., Gayet, J. F., Krämer, M., Schneider, J., Borrmann, S., Schmale, J., Jessberger, P., Hamburger, T., Lichtenstern, M., Scheibe, M., Gourbeyre, C., Meyer, J., Kübbeler, M., Frey, W., Kalesse, H., Butler, T., Lawrence, M. G., Holzäpfel, F., Arnold, F., Wendisch, M., Döpelheuer, A., Gottschaldt, K., Baumann, R., Zöger, M., Sölch, I., Rautenhaus, M., and Dörnbrack, A.: In - situ observations of young contrails - overview and selected results from the CONCERT campaign, Atmos. Chem. Phys., 10, 9039 – 9056, 2010.

Voigt, C., Schumann, U., Minikin, A., Abdelmonem, A., Afchine, A., Borrmann, S., Boettcher, M., Buchholz, B., Bugliaro, L., Costa, A., Curtius, J., Dollner, M., Dörnbrack, A., Dreiling, V., Ebert, V., Ehrlich, A., Fix, A., Forster, L., Frank, F., Fütterer, D., Giez, A., Graf, K., Grooß, J. U., Groß, S., Heimerl, K., Heinold, B., Hüneke, T., Järvinen, E., Jurkat, T., Kaufmann, S., Kenntner, M., Klingebiel, M., Klimach, T., Kohl, R., Krämer, M., Krisna, T. C., Luebke, A., Mayer, B., Mertes, S., Molleker, S., Petzold, A., Pfeilsticker, K., Port, M., Rapp, M., Reutter, P., Rolf, C., Rose, D., Sauer, D., Schäfler, A., Schlage, R., Schnaiter, M., Schneider, J., Spelten, N., Spichtinger, P., Stock, P., Walser, A., Weigel, R., Weinzierl, B., Wendisch, M., Werner, F., Wernli, H., Wirth, M., Zahn, A., Ziereis, H., and Zöger, M.: ML - CIRRUS: The Airborne Experiment on Natural Cirrus and Contrail Cirrus with the High - Altitude Long - Range Research Aircraft HALO, Bull. Amer. Meteor. Soc., 98, 271 – 288, 2017.

Wernli, H., Boettcher, M., Joos, H., Miltenberger, A. K., and Spichtinger, P.: A trajectory - based classification of ERA - Interim ice clouds in the region of the North Atlantic storm track, Geophys. Res. Lett., 43, 6657 – 6664, 2016.

Wielicki, B. A., Barkstrom, B. R., Harrison, E. F., Lee, R. B., Smith, G. L., and Cooper, J. E.: Clouds and the Earth's Radiant Energy System (CERES): An Earth Observing System Experiment, Bull. Amer. Meteor. Soc., 77, 853 – 868, 1996.

Wild, M., Folini, D., Hakuba, M. Z., Schär, C., Seneviratne, S. I., Kato, S., Rutan, D., Ammann, C., Wood, E. F., and König Langlo, G.: The energy balance over land and oceans: an assessment based on direct observations and CMIP5 climate models, Clim. Dynam., 44, 3393 – 3429, 2015.

Winker, D. M., Hunt, W. H., and McGill, M. J.: Initial performance assessment of CALIOP, Geophys. Res. Lett., 34, 2007.

Winker, D. M., Vaughan, M. A., Omar, A., Hu, Y., Powell, K. A., Liu, Z., Hunt, W. H., and Young, S. A.: Overview of the CALIPSO Mission and CALIOP Data Processing Algorithms Journal of Atmospheric and Oceanic Technology, J. Atmos. Ocean. Technol., 26, 2310 – 2323, 2009.

WMO: Meteorology Three - Dimensional Science: Second Session of the Commission for Aerology, WMO Bulletin, IV, 134 – 138, 1957.

Wofsy, S. C.: HIAPER Pole - to - Pole Observations (HIPPO): fine - grained, global - scale measurements of climatically important atmospheric gases and aerosols, Philos. Trans. Roy. Soc. London, 369, 2073 – 2086, 2011.

Wu, C., Liu, X., Diao, M., Zhang, K., Gettelman, A., Lu, Z., Penner, J. E., and Lin, Z.: Direct comparisons of ice cloud macro - and microphysical properties simulated by the Community Atmosphere Model Version 5 with HIPPO aircraft observations, Atmos. Chem. Phys., 17, 4731 – 4749, 2017.

Zängl, G., Reinert, D., Rípodas, P., and Baldauf, M.: The ICON (ICOsahedral Non - hydrostatic) modelling framework of DWD and MPI - M: Description of the non - hydrostatic dynamical core, Quart. J. Roy. Meteorol. Soc., 141, 563 – 579, 2015.

Zhou, C., Penner, J. E., Lin, G., Liu, X., and Wang, M.: What controls the low ice number concentration in the upper troposphere?, Atmos. Chem. Phys., 16, 12411 – 12424, 2016.

Zou, B., Buxi, G. S., and Hansen, M.: Optimal 4 - D Aircraft Trajectories in a Contrail - sensitive Environment, Networks and Spatial Economics, 16, 415 – 446, 2016.

A Constants used for hydrometeor classes

Table A.1: Coefficients for the mass-size relation (Eg. 2.7) for all hydrometeors. Note that these value are valid for m in kg and L in μm.

	a_{geo}	b_{geo}	ν	μ	m_{min}	m_{max}
cloud	0.124	0.3	1	1	4.2×10^{-15}	2.6×10^{-10}
ice	0.835	0.39	0	0.3	10^{-12}	10^{-6}
rain	0.124	0.3	0	0.3	2.6×10^{-10}	3×10^{-6}
snow	5.13	0.5	0	0.5	10^{-10}	2×10^{-5}
graupel	0.142	0.314	1	0.3	10^{-9}	5×10^{-4}
hail	0.137	0.3	1	0.3	2.6×10^{-9}	5×10^{-4}

Table A.2: Coefficients for the mass-velocity relation (Eg. 2.16) for all hydrometeors. Note that these value are valid for m in kg and v in m s^{-1}.

	a_{vel}	b_{vel}	v_{min}	v_{max}
cloud	3.75×10^{5}	0.667	0.0	1
ice	27.7	0.216	0.0	3
rain	114.014	0.234	0.1	20
snow	8.294	0.125	0.1	3
graupel	86.894	0.268	0.1	10
hail	39.3	0.167	0.1	30

Table A.3: Constants characterizing the natural and the contrail ice class.

Symbol	Unit	Natural ice class		Contrail ice class	
		Value	Reference	Value	Reference
a_{geo}	-	0.835	A. Seifert, pers. comm., 01 June, 2017	0.124 , 4.25	Spichtinger and Gierens (2009) Heymsfield and Iaquinta (2000)
b_{geo}	-	0.39	A. Seifert, pers. comm., 01 June, 2017	0.33, 0.45	Spichtinger and Gierens (2009) Heymsfield and Iaquinta (2000)
m_{min}	kg	10^{-12}	Seifert and Beheng (2006)	10^{-15}	Gruber et al. (2018)
m_{max}	kg	10^{-6}	Seifert and Beheng (2006)	2×10^{-11}	Gruber et al. (2018)
L_{min}	μm	17.5		1.24	
L_{max}	μm	3800		58	
μ		0.333 Seifert and Beheng (2006)			
ν		1 Seifert and Beheng (2006)			
a_{vel}		317.0 Seifert and Beheng (2006)			
b_{vel}		0.363 Seifert and Beheng (2006)			

B Calculation of the cloud optical properties

Spectral wavelength bands used in GRAALS and RRTM

Table B.1: Spectral wavelength bands (in μm) used in RRTM (Mlawer et al., 1997).

solar		thermal	
1	3.85 - 3.1	15	1000.0 - 28.57
2	3.08 - 2.5	16	28.57 - 20.0
3	2.5 - 2.15	17	20.0 - 15.87
4	2.15 - 1.94	18	15.87 - 14.29
5	1.94 - 1.63	19	14.29 - 12.2
6	1.63 - 1.3	20	12.2 - 10.2
7	1.3 - 1.24	21	10.2 - 9.26
8	1.24 - 0.78	22	9.26 - 8.47
9	0.78 - 0.63	23	8.47 - 7.19
10	0.63 - 0.44	24	7.19 - 6.76
11	0.44 - 0.34	25	6.76 - 5.56
12	0.34 - 0.26	26	5.56 - 4.81
13	0.26 - 0.2	27	4.81 - 4.44
14	12.2 - 3.85	28	4.44 - 4.2
		29	4.2 - 3.85
		30	3.85 - 3.08

Table B.2: Spectral wavelength bands (in μm) used in GRAALS (Ritter and Geleyn, 1992).

	solar		thermal
1	1.53 - 4.64	4	12.5 - 20.0
2	0.7 - 1.53	5	8.33 - 9.01
3	0.25 - 0.7	6	9.01 - 10.31
		7	4.64 - 8.33

Effective radius

Here, the conversion from $r_{\mathrm{e}}(L)$ to $r_{\mathrm{e}}(m)$ for using the parameterization after Fu et al. (1998) with the two moment microphysical scheme (Seifert and Beheng, 2006) is described.
The formulation for effective radius r_{e} after Fu et al. (1998), assuming ice crystals being hexagonal columns with diameter D and length L reads as:

$$r_{\mathrm{e}} = \frac{1}{2} \frac{\int_0^\infty D^2 L f(L)\,\mathrm{d}L}{\int_0^\infty f(L)\left(DL + \frac{\sqrt{3}}{4} D^2\right)\mathrm{d}L} \qquad \text{(B.1)}$$

Partial moments of the generalized Γ-distribution as $f(m)$ as defined in Eq. 2.2 satisfy Eq. B.2.

$$\int_0^\infty f^k f(m)\,\mathrm{d}m = \frac{A}{\mu}\Gamma\left(\frac{\nu+k+1}{\mu}\right)\lambda^{-\frac{\nu+k+1}{\mu}} \qquad \text{(B.2)}$$

With coefficients A and λ defined in Eq. 2.5 and Eq. 2.6.
However, in the parameterization after Fu et al. (1998) the PSD f is assumed to be dependent from particle size L. Later, to be consistent to the convention used in the microphysics, $f(L)$ will be transformed to $f(m)$, hence a transformation

of the parameters A, λ, μ and ν is required. Until further notice, in the following manipulations, when μ and ν is written, implicitly μ_{L} and ν_{L} are assumed.
Transformation relations as defined in Eq. B.3 are used for converting $f(m)$ to $f(L)$ and vice versa.

$$\begin{aligned}
\nu_{\mathrm{L}} &= \frac{\nu_{\mathrm{m}}+1}{b_{\mathrm{geo}}} - 1 \\
\mu_{\mathrm{L}} &= \frac{\mu_{\mathrm{m}}}{b_{\mathrm{geo}}} \\
\lambda_{\mathrm{L}} &= \lambda_{\mathrm{m}} a_{\mathrm{geo}}^{-\frac{\mu_{\mathrm{m}}}{b_{\mathrm{geo}}}} \\
A_{\mathrm{L}} &= \frac{A_{\mathrm{m}}}{b_{\mathrm{geo}}} a_{\mathrm{geo}}^{-\frac{1}{b_{\mathrm{geo}}}(\nu_{\mathrm{X}}+1)}
\end{aligned} \tag{B.3}$$

For the following derivations, the geometry of ice crystals is specified more exactly. In the preceding section describing the basics of the microphysical processes (section 2.4), a power-law was introduced to connect diameter D and mass m of particles (Eq. 2.7) with coefficients a_{geo} and b_{geo} representing the particles' shape. Here, D represents the average size rather than a geometric diameter. This assumption holds when considering an ensemble of randomly orientated particles. However, as ice crystals are assumed to be hexagonal columns, two possible dimensions exist characterizing the size such particles, i. e. the length of the column L and the diameter D of the base. As with increasing ice crystal mass, the length of the column increases stronger than the diameter of its base, Eq. 2.7 can be considered as a relation of column length L depending on particle mass m (Eq. B.4).

$$L = a_{\mathrm{geo}} m^{\mathrm{b_{geo}}} \tag{B.4}$$

Several geometric properties a hexagonal column are necessary in the following. The area O of an hexagon is defined in Eq. B.5.

$$O = D^2 \frac{3\sqrt{3}}{8} \tag{B.5}$$

The volume V of an hexagonal column is given in Eq. B.6.

$$V = OL = D^2 L \frac{3\sqrt{3}}{8} \tag{B.6}$$

Furthermore, the surface area S of an hexagonal column follows Eq. B.7.

$$S = 2O + 6\frac{D}{2}L = 3\left(\frac{\sqrt{3}}{4}D^2 L + DL\right) \tag{B.7}$$

The density ρ of a single particle simply follows Eq. B.8. s

$$\rho = \frac{m}{V} \tag{B.8}$$

As D and L are not independent from each other, $D(L)$ is assembled by combining Eq.B.4, Eq.B.6 and Eq.B.8.

$$D(L) = pL^{\frac{1-b_{\text{geo}}}{2b_{\text{geo}}}} \; ; \; p = \sqrt{\frac{8}{3\sqrt{3}\rho a_{\text{geo}}{}^{\frac{1}{b_{\text{geo}}}}}} \tag{B.9}$$

Applying Eq. B.9 to Eq. B.1 yields Eq. B.10.

$$r_{\mathrm{e}} = \frac{1}{2} \frac{\int_0^\infty p^2 L^{\frac{1}{b_{\mathrm{geo}}}} f(L)\, \mathrm{d}L}{\int_0^\infty \left(pL^{\frac{1}{2b_{\mathrm{geo}}}+\frac{1}{2}} + \frac{\sqrt{3}}{4} p^2 L^{\frac{1}{b_{\mathrm{geo}}}-1} \right) f(L)\, \mathrm{d}L} \tag{B.10}$$

Integrating Eq. B.10 using the identity defined in Eq. B.2 yields Eq. B.11.

$$r_{\mathrm{e}} = \frac{p\frac{A}{\mu}\Gamma(\psi_1)\,\lambda^{-\psi_1} + \frac{A}{\mu}\Gamma(\psi_2)\,\lambda^{-\psi_2}}{\frac{\sqrt{3}}{4} p\frac{A}{\mu}\Gamma(\psi_3)\,\lambda^{-\psi_3}} \tag{B.11}$$

With coefficients ψ_1 to ψ_3 given in Eq. B.12.

$$\begin{aligned} \psi_1 &= \frac{\nu - 2}{\mu} \\ \psi_2 &= \frac{\nu + \frac{3}{2} + \frac{b_{\mathrm{geo}}}{2}}{\mu} \\ \psi_3 &= \frac{\nu + 2 - b_{\mathrm{geo}}}{\mu} \end{aligned} \tag{B.12}$$

The transformation rules defined in Eq. B.3 are applied to Eq. B.11. Inserting expressions for A and λ according to Eq. 2.5 and Eq. 2.5 gives Eq. B.13.

$$r_{\mathrm{e}} = \frac{1}{2} \left(c_1 \overline{m}^{c_2} + c_3 \overline{m}^{c_4} \right)^{-1} \tag{B.13}$$

With coefficients c_1 to c_4 given in Eq. B.14.

$$
\begin{aligned}
c_1 &= \sqrt{\frac{3\sqrt{3}\rho a_{\text{geo}}}{8} \frac{\Gamma\left(\frac{\nu+\frac{3}{2}-\frac{b_{\text{geo}}}{2}}{\mu}\right)}{\Gamma\left(\frac{\nu+2}{\mu}\right)} \left[\frac{\Gamma\left(\frac{\nu+1}{\mu}\right)}{\Gamma\left(\frac{\nu+2}{\mu}\right)}\right]^{\frac{1}{2}\left(1-b_{\text{geo}}\right)}} \\
c_2 &= \frac{1}{2}\left(1-b_{\text{geo}}\right) \\
c_3 &= \frac{\sqrt{3}}{4a_{\text{geo}}} \frac{\Gamma\left(\frac{\nu+2-b_{\text{geo}}}{\mu}\right)}{\Gamma\left(\frac{\nu+2}{\mu}\right)} \left[\frac{\Gamma\left(\frac{\nu+1}{\mu}\right)}{\Gamma\left(\frac{\nu+2}{\mu}\right)}\right]^{b_{\text{geo}}} \\
c_4 &= b_{\text{geo}}
\end{aligned}
\tag{B.14}
$$

Mean axis ratio

According to Fu (2007), the asymmetry parameter g is parameterized as function of the crystals' mean axis ratio A_{r}. As for the effective radius, $A_{\text{r}}(L)$ needs to be transformed to $A_{\text{r}}(m)$ to be used with the two moment microphysical scheme after Seifert and Beheng (2006). Also here, the particles are assumed to be of a hexagonal columnar shape with diameter D and length L.

The mean axis ratio A_{r} is defined as:

$$
A_{\text{r}} = \frac{\int_0^\infty \frac{D}{L} A_{\text{p}} f(L)\,\mathrm{d}L}{\int_0^\infty A_{\text{p}} f(L)\,\mathrm{d}L} \tag{B.15}
$$

Here, $\mathrm{A_p}$ is the statistical average projected area of a particle that is $1/4$ of its total surface area when assuming random orientation.

$$A_{\mathrm{p}} = \frac{1}{4}S = \frac{3}{4}\left(\frac{\sqrt{3}}{4}D^2L + DL\right) \tag{B.16}$$

For converting $A_{\mathrm{r}}(L)$ to $A_{\mathrm{r}}(m)$, the same approach is used as described above. Applying Eq. B.9 and Eq. B.16 to Eq. B.15 yields Eq. B.17.

$$A_{\mathrm{r}} = \frac{\int_0^\infty \left(\frac{\sqrt{3}}{4}p^2L^{\frac{3}{2}\left(\frac{1}{b_{\mathrm{geo}}}-1\right)} + L^{\frac{1}{2}\left(\frac{1}{b_{\mathrm{geo}}}-1\right)}\right) f(L)\,\mathrm{d}L}{\int_0^\infty \left(\frac{\sqrt{3}}{4}pL^{\frac{1}{b_{\mathrm{geo}}}} + L^{\frac{1}{2}\left(\frac{1}{b_{\mathrm{geo}}}+1\right)}\right) f(L)\,\mathrm{d}L} \tag{B.17}$$

Integrating Eq. B.17 using the identity defined in Eq. B.2 yields Eq. B.18.

$$A_{\mathrm{r}} = \frac{\frac{\sqrt{3}}{4}p^2\frac{A}{\mu}\Gamma(\psi_1)\lambda^{-\psi_1} + \frac{A}{\mu}\Gamma(\psi_2)\lambda^{-\psi_2}}{\frac{\sqrt{3}}{4}p\frac{A}{\mu}\Gamma(\psi_3)\lambda^{-\psi_3} + \frac{A}{\mu}\Gamma(\psi_4)\lambda^{-\psi_4}} \tag{B.18}$$

With coefficients ψ_1 to ψ_4 given in Eq. B.19.

$$\begin{aligned}
\psi_1 &= \frac{\nu - \frac{1}{2} + \frac{3}{2b_{\mathrm{geo}}}}{\mu} \\
\psi_2 &= \frac{\nu + \frac{1}{2} + \frac{1}{2b_{\mathrm{geo}}}}{\mu} \\
\psi_3 &= \frac{\nu + 1 + \frac{1}{b_{\mathrm{geo}}}}{\mu} \\
\psi_4 &= \frac{\nu + \frac{3}{2} + \frac{1}{2b_{\mathrm{geo}}}}{\mu}
\end{aligned} \tag{B.19}$$

Now, the transformation rules defined in Eq. B.3 are applied to Eq. B.18. Inserting expressions for A and λ according to Eq. 2.5 and Eq. 2.5 gives Eq. B.20.

$$A_{\mathrm{r}} = \frac{c_1 \overline{m}^{c_2} + c_3 \overline{m}^{c_4}}{c_5 \overline{m}^{c_6} + c_7 \overline{m}^{c_8}} \tag{B.20}$$

With coefficients c_1 to c_8 given in Eq. B.21.

$$\begin{aligned}
c_1 &= \frac{2}{3\rho} a_{\mathrm{geo}}{}^{-\frac{3}{2}} \Gamma\left(\frac{\nu + \frac{5}{2} - \frac{3b_{\mathrm{geo}}}{2}}{\mu}\right) \left[\frac{\Gamma\left(\frac{\nu+1}{\mu}\right)}{\Gamma\left(\frac{\nu+2}{\mu}\right)}\right]^{-\frac{1}{2}\left(5-3b_{\mathrm{geo}}\right)} \\
c_2 &= -\frac{1}{2}\left(5 - 3b_{\mathrm{geo}}\right) \\
c_3 &= \frac{1}{\sqrt{a_{\mathrm{geo}}}} \Gamma\left(\frac{\nu + \frac{3}{2} - \frac{b_{\mathrm{geo}}}{2}}{\mu}\right) \left[\frac{\Gamma\left(\frac{\nu+1}{\mu}\right)}{\Gamma\left(\frac{\nu+2}{\mu}\right)}\right]^{-\frac{1}{2}\left(3-b_{\mathrm{geo}}\right)} \\
c_4 &= -\frac{1}{2}\left(3 - b_{\mathrm{geo}}\right) \\
c_5 &= \frac{1}{\sqrt{2\sqrt{3}\rho}} \Gamma\left(\frac{\nu+2}{\mu}\right) \left[\frac{\Gamma\left(\frac{\nu+1}{\mu}\right)}{\Gamma\left(\frac{\nu+2}{\mu}\right)}\right]^{-2} \\
c_6 &= -2 \\
c_7 &= \sqrt{a_{\mathrm{geo}}} \Gamma\left(\frac{\nu + \frac{3}{2} + \frac{b_{\mathrm{geo}}}{2}}{\mu}\right) \left[\frac{\Gamma\left(\frac{\nu+1}{\mu}\right)}{\Gamma\left(\frac{\nu+2}{\mu}\right)}\right]^{-\frac{1}{2}\left(3+b_{\mathrm{geo}}\right)} \\
c_8 &= -\frac{1}{2}\left(3 + b_{\mathrm{geo}}\right)
\end{aligned} \tag{B.21}$$

Spectral averaging to wide bands

In this section, the averaging procedure is exemplary expanded for β'_{ext}. The derivations are printed with friendly permission of U. Blahak, H. Muskatel, and P. Khain.

For β'_{ext}, the 'correct' definition equation for spectral averaging would be as follows, considering the Planck-function S and a certain PDF of the optical thickness, composed of a PDF of *IWC* ρ_{c} and a range of model level thicknesses Δz:

$$\begin{aligned}
&\exp\left[-\overline{\overline{\overline{\beta'_{\mathrm{ext}}(r_{\mathrm{e}})}^{\lambda}}}^{\rho_c\Delta z}\overline{\rho_c\Delta z}^{?}\right] \\
&\stackrel{!}{=}\frac{\int_0^\infty\int_0^\infty\int_{\lambda_{min}}^{\lambda_{min}} f(\rho_c)\,g(\Delta z)\,S(\lambda)\exp\left[-\beta'_{\mathrm{exp}}(r_{\mathrm{e}},\lambda)\,\rho_c\Delta z\right]\mathrm{d}\lambda\,\mathrm{d}\rho_c\,\mathrm{d}\Delta z}{\int_0^\infty\int_0^\infty\int_{\lambda_{min}}^{\lambda_{min}} f(\rho_c)\,g(\Delta z)\,S(\lambda)\,\mathrm{d}\lambda\,\mathrm{d}\rho_c\,\mathrm{d}\Delta z}
\end{aligned} \tag{B.22}$$

However, due to the unclear definition of $\overline{\rho_c\Delta z}^{?}$, averaging over the wavelength is performed separately. Afterwards, the wavelength-averaged $\overline{\beta'_{\mathrm{ext}}}^{\lambda}$ is averaged over ρ_c and Δz.

Wavelength average:

$$\begin{aligned}
&\exp\left[-\overline{\beta'_{\mathrm{ext}}(r_{\mathrm{e}})}^{\lambda}\rho_c\Delta z\right] \\
&\stackrel{!}{=}\frac{\int_{\lambda_{min}}^{\lambda_{min}} S(\lambda)\exp\left[-\beta'_{\mathrm{exp}}(r_{\mathrm{e}},\lambda)\,\rho_c\Delta z\right]\mathrm{d}\lambda}{\int_{\lambda_{min}}^{\lambda_{min}} S(\lambda)\,\mathrm{d}\lambda} \\
&\longrightarrow \overline{\beta'_{\mathrm{ext}}}^{\lambda}(r_{\mathrm{e}},\rho_c,\Delta z) \\
&=-\frac{1}{\rho_c\Delta z}\ln\left[\frac{\int_{\lambda_{min}}^{\lambda_{min}} S(\lambda)\exp\left[-\beta'_{\mathrm{exp}}(r_{\mathrm{e}},\lambda)\,\rho_c\Delta z\right]\mathrm{d}\lambda}{\int_{\lambda_{min}}^{\lambda_{min}} S(\lambda)\,\mathrm{d}\lambda}\right]
\end{aligned} \tag{B.23}$$

For numerical computations, the trapezoidal rule is employed on the non-equidistant original wavelength discretization of the single-particle scattering data.

ρ_c and Δz average:

$$\overline{\overline{\overline{\beta'_{\text{ext}}}^{\lambda}}^{\rho_c}}^{\Delta z}(r_{\text{e}}) = \frac{\int_0^\infty \int_0^\infty f(\rho_c)\, g(\Delta z)\, \overline{\beta'_{\text{ext}}}^{\lambda}(r_{\text{e}}, \rho_c, \Delta z)\, \mathrm{d}\rho_c \mathrm{d}\Delta z}{\int_0^\infty \int_0^\infty f(\rho_c)\, g(\Delta z)\, \mathrm{d}\rho_c \mathrm{d}\Delta z} \tag{B.24}$$

An equidistant discretization is used, and again the trapezoidal rule employed.

$$\begin{aligned}\overline{\overline{\overline{\beta'_{\text{ext}}}^{\lambda}}^{\rho_c}}^{\Delta z}(r_{\text{e}}) &\approx \frac{\sum_1^{N_\Delta} \sum_1^{N_\rho} f(\rho_{c,k})\, g(\Delta z_i)\, \overline{\beta'_{\text{ext}}}^{\lambda}(r_{\text{e}}, \rho_{c,k}, \Delta z_i)\, \Delta\rho_{c,k} \Delta(\Delta z)_i}{\sum_1^{N_\Delta} \sum_1^{N_\rho} f(\rho_{c,k})\, g(\Delta z_i)\, \Delta\rho_{c,k} \Delta(\Delta z)_i} \\ &= \frac{\sum_1^{N_\Delta} \sum_1^{N_\rho} f(\rho_{c,k})\, g(\Delta z_i)\, \overline{\beta'_{\text{ext}}}^{\lambda}(r_{\text{e}}, \rho_{c,k}, \Delta z_i)}{\sum_1^{N_\Delta} \sum_1^{N_\rho} f(\rho_{c,k})\, g(\Delta z_i)}\end{aligned} \tag{B.25}$$

Here, f is a triangular distribution on the interval $[0, \rho_{c,max}]$:

$$\mathrm{f}(\rho_{c,k}) = \begin{cases} \frac{2}{\rho_{c,max}} - \frac{2}{\rho_{c,max}^2}\rho_c & 0 \leq \rho_{c,k} \leq \rho_{c,max} \\ 0 & \text{else} \end{cases} \tag{B.26}$$

$$\rho_{c,k} = \left(k - \frac{1}{2}\right)\Delta\rho_c,\ k = 1, \ldots N_p,\ N_p = \frac{\rho_{c,max}}{\Delta\rho_c} \tag{B.27}$$

Here, g is defined as a uniform distribution on the interval $[\Delta z_{min}, \Delta z_{max}]$:

$$g(\Delta z_i) = \begin{cases} \frac{1}{\Delta z_{max} - \Delta z_{min}} & \Delta z_{min} \leq \Delta z_i \leq \Delta z_{max} \\ 0 & \text{else} \end{cases} \tag{B.28}$$

$$\Delta z_i = \Delta z_{min} + \left(i - \frac{1}{2}\right)\Delta(\Delta z),\ i = 1, \ldots N_\Delta,\ N_\Delta = \frac{\Delta z_{max} - \Delta z_{min}}{\Delta(\Delta z)} \tag{B.29}$$

Here, the parameters are chosen as $\rho_{\mathrm{c,max}} = 1\ \mathrm{g\ m^{-3}}$, $\Delta z_{\mathrm{min}} = 100$ m and $\Delta z_{\mathrm{max}} = 600$ m.

C Spectral wavelength bands used in GRAALS and RRTM

Table C.1: Coefficients for extinction coefficient β_{ext} in Eq. 2.73 for GRAALS wavelength bands.

	c_0	c_1	c_2	c_3	c_4	c_5	c_6
1	540020.000	-58078.5016	2575.72361	878.104246	138.038190	-20.7103434	1.00240512
2	60879.7019	-17192.0924	1943.60413	3.04056517	23.4421437	-6.68895412	0.76926125
3	278490.000	-39036.0991	1693.63518	20.4268802	107.265599	-15.3363761	0.67062524
4	17709.7195	222.317012	30.0150536	489.687169	-8.97471779	0.20477665	0.01100354
5	19038.0756	-769.751515	301.951941	200.599468	-14.8225136	0.21218129	0.10931606
6	71274.7235	2463.55273	176.148837	195.684728	26.6328974	0.94208657	0.06858052
7	28094.8904	720.972625	62.0364227	277.449875	-13.4742736	0.74451826	0.02150099
8	63642.9792	-6517.99425	1099.85698	243.253340	-15.9974340	-1.16346252	0.40432223

Table C.2: Coefficients for single scattering albedo ω in Eq. 2.74 for GRAALS wavelength bands.

	c_0	c_1	c_2	c_3	c_4	c_5	c_6	c_7
1	1503.00519	101.654637	0.80341886	0.00023585	1567.02090	115.201101	1.11886384	0.00079132
2	515.878591	6.49639138	0.08276779	0.00004856	515.906105	6.56062545	0.08370302	0.00005335
3	0.20190304	-143.67240	34.8842452	-0.00007513	0.20194326	-143.672381	34.8842313	-0.00007162
4	-616.459213	288.510394	-1.55401854	0.16046003	965.072897	289.058726	-1.03984205	0.29577464
5	-862.909994	369.632956	-24.3016266	0.73276986	-749.095146	544.311870	-39.9292381	1.31879741
6	24.5030815	-56.6597847	22.4902792	0.02663082	156.739985	31.8715473	41.6763200	0.05000000
7	-74.9801800	32.3838700	-1.32676000	0.04691000	-76.8780700	45.9688400	-2.33181300	0.08864782
8	-498.331209	1384.13888	-53.3837983	6.34715235	928.888203	1622.55919	-91.2200371	11.8833432

Table C.3: Coefficients for asymmetry parameter g in Eq. 2.76 for GRAALS wavelength bands.

	c_0	c_1	c_2	c_3	c_4	c_5	c_6
1	0.89333805	-8.63761743	21.5219918	-3.25967766	0.95354260	-9.05508000	21.4585469
2	0.72575541	-1.35360211	0.74832965	-0.11047952	0.82863973	-1.37886149	0.56378475
3	0.08490004	2.03362664	-1.27662950	0.34652694	0.11276958	2.27793248	-0.76577231
4	1.44444011	-5.79090754	6.49972307	-1.20428285	1.45101930	-5.49046051	5.28305024
5	0.86935869	-1.86868592	1.11311759	-0.06285900	0.91767959	-1.92823314	1.07235770
6	0.88967491	-1.62682053	0.77536951	-0.01304385	0.92783977	-1.68162695	0.78164272
7	0.86342860	-1.56231382	0.75648254	-0.00834041	0.87766114	-1.57081942	0.74669678
8	0.85220449	-1.75466490	0.95126199	-0.02191880	0.87505573	-1.78421632	0.93940595

Table C.4: Coefficients for extinction coefficient β_{ext} in Eq. 2.77 for GRAALS wavelength bands.

	$c_{1,0}$	$c_{1,1}$	$c_{1,2}$	$c_{2,0}$	$c_{2,1}$	$c_{2,2}$	$c_{3,0}$	$c_{3,1}$	$c_{3,2}$
1	2.25651225e+03	-1.18200363e+00	1.85876391e+01	1.93521777e+03	-1.07056981e+00	2.01486248e+00	1.89490365e+03	-1.05319405e+00	0
2	1.86410201e+03	-1.08712336e+00	7.27453246e+00	1.74656230e+03	-1.04362183e+00	1.19587445e+00	1.79205315e+03	-1.04209999e+00	0
3	1.69181603e+03	-1.04463256e+00	3.08062592e+00	1.65347784e+03	-1.02739040e+00	7.99955802e-01	1.73788689e+03	-1.03645435e+00	0
4	3.06599650e+05	7.16096801e-05	-3.06526626e+05	6.74912209e+05	-1.21406634e-04	-6.74571840e+05	2.81835046e+03	-1.12043012e+00	0
5	8.37129973e+05	-1.18388602e-04	-8.36746626e+05	1.00701104e+03	-7.24080462e-01	-2.94367855e+01	2.17029592e+03	-1.07118149e+00	0
6	-2.66375660e-03	4.03989422e+00	2.06537195e+02	2.47250409e+03	-1.14285816e+00	5.43005969e+00	2.25095287e+03	-1.08438660e+00	0
7	-2.22599856e-06	6.58432531e+00	1.95925581e+02	3.57709225e+03	-1.22369076e+00	1.22795013e+00	2.47267932e+03	-1.10598151e+00	0
8	-3.58965590e+00	1.70778763e+00	3.90986067e+02	3.37380122e+03	-1.26127491e+00	9.84848914e+00	2.23347618e+03	-1.08400166e+00	0

Table C.5: Coefficients for single scattering albedo ω in Eq. 2.77 for GRAALS wavelength bands.

	$c_{1,0}$	$c_{1,1}$	$c_{1,2}$	$c_{2,0}$	$c_{2,1}$	$c_{2,2}$	$c_{3,0}$	$c_{3,1}$	$c_{3,2}$
1	-1.77484297e-01	-3.82659292e-01	1.57411270e-01	-3.99713397e+03	-9.35025282e-06	3.99712987e+03	-3.90469846e+03	-1.36180240e-05	3.90464094e+03
2	-3.95088467e+02	-1.38803637e-06	3.95088075e+02	-4.69627738e+02	-3.20613944e-06	4.69624928e+02	-5.83150345e+02	-5.26289656e-06	5.83142203e+02
3	-2.00411661e+00	-5.53000038e-07	2.00411587e+00	-3.74264515e+00	-9.17983686e-07	3.74263852e+00	-4.00552597e+00	-1.69134430e-06	4.00550801e+00
4	1.10545624e+00	-9.50675210e-01	4.51590386e-01	8.12808997e-01	-1.09473634e+00	5.00557021e-01	-1.19184186e-01	1.73105279e-01	7.34958637e-01
5	8.80652139e-01	-8.14816312e-01	4.45618477e-01	5.29896829e-01	-5.37977174e-01	4.21587773e-01	-3.13798540e+03	1.22089852e-05	3.13862082e+03
6	-7.60878380e+03	1.12738374e-05	7.60940910e+03	-9.42787209e-01	-8.13821122e-01	5.44872162e-01	-1.21627107e+07	-6.09496567e+00	4.97988214e-01
7	4.81469213e-14	1.08360561e+01	2.72996507e-01	-3.50807690e+00	-9.83284687e-01	5.91069272e-01	-4.74763482e+04	-4.22223945e+00	4.90639909e-01
8	1.18094219e-03	1.96103353e+00	1.73555077e-01	-1.23069431e+01	-1.73041492e+00	4.89368484e-01	-1.52202307e+02	-2.55364031e+00	4.80018143e-01

Table C.6: Coefficients for asymmetry parameter g in Eq. 2.77 for GRAALS wavelength bands.

	$c_{1,0}$	$c_{1,1}$	$c_{1,2}$	$c_{2,0}$	$c_{2,1}$	$c_{2,2}$	$c_{3,0}$	$c_{3,1}$	$c_{3,2}$
1	1.35155139e-03	1.37919662e+00	8.61329693e-01	-5.21025320e-01	-1.07539351e+00	9.38249930e-01	-2.01780162e-01	-5.79270612e-01	9.53270216e-01
2	-4.08189497e-01	-1.27270997e+00	8.71600615e-01	-2.16114961e-01	-6.48494355e-01	8.97293947e-01	-2.38862952e-01	-6.92283725e-01	8.95753046e-01
3	-9.68604617e-02	-5.33538940e-01	8.92138933e-01	-1.57506916e-01	-9.68044744e-01	8.81523330e-01	-1.71271436e-01	-9.69125115e-01	8.81838376e-01
4	-1.88135801e+00	-6.44478276e-01	1.11637275e+00	-3.31316873e+00	-1.10913884e+00	9.45509280e-01	-9.47504022e+00	-1.49199962e+00	9.28185824e-01
5	-2.41189309e+00	-1.33289366e+00	9.58865701e-01	-2.75897852e+00	-1.44231454e+00	9.47403178e-01	-3.41318673e+00	-1.51533707e+00	9.46533750e-01
6	-2.14454970e+00	-1.63554204e+00	9.62747659e-01	-3.16371321e-01	-2.30272653e-01	1.10433012e+00	-4.81294210e+00	-1.55463251e+00	9.83758483e-01
7	-2.06515711e+00	-1.95356112e+00	9.17550008e-01	1.04458555e-03	1.22456891e+00	8.76228385e-01	-3.00488637e+01	-1.96229250e+00	9.79789555e-01
8	-2.32150720e+00	-2.84391926e+00	8.84933671e-01	-6.57145274e-01	-2.78203089e-01	1.20223190e+00	-3.36755340e+01	-2.08919516e+00	9.74027070e-01

Table C.7: Coefficients for extinction coefficient β_{ext} in Eq. 2.73 for RRTM wavelength bands.

	c_0	c_1	c_2	c_3	c_4	c_5	c_6
1	4.734484e+04	1.225170e+03	2.732055e+01	1.760128e+03	-3.976866e+01	7.951256e-01	9.697505e-03
2	1.527138e+04	1.738261e+03	1.203636e+01	4.511182e+02	-2.378510e+01	1.030221e+00	3.495695e-03
3	2.922573e+04	2.215985e+03	1.147692e+02	3.494118e+02	-2.919974e+01	1.699924e+00	3.833563e-02
4	5.316819e+04	7.570687e+03	7.097043e+02	4.273058e+02	-3.393479e+01	3.646986e+00	2.646141e-01
5	-1.659143e+04	1.398913e+03	8.840784e+02	-5.498606e+01	1.457568e+01	-1.062198e+00	3.555384e-01
6	2.429853e+04	-5.736491e+03	1.665388e+03	6.390146e+01	2.438907e+01	-2.078254e+00	6.410155e-01
7	3.924131e+04	2.727987e+03	9.062342e+01	4.288317e+02	-1.987242e+01	1.629829e+00	3.246777e-02
8	3.090728e+04	-7.571926e+01	9.127094e+01	2.183164e+02	-1.294156e+01	5.771626e-01	3.105830e-02
9	3.657389e+06	-7.220574e+04	3.269946e+04	2.068559e+04	-1.515086e+03	5.045349e+01	1.188795e+01
10	2.767019e+04	-1.970338e+03	4.555923e+02	1.025239e+02	-7.482968e+00	-1.661276e-01	1.673306e-01
11	3.359162e+04	-1.828578e+03	6.643632e+02	1.105129e+02	-7.839641e+00	-2.997420e-02	2.469626e-01
12	6.972496e+04	-1.153168e+04	1.789593e+03	1.151387e+02	4.684693e+00	-3.902173e+00	6.735007e-01
13	1.879671e+05	-6.078889e+04	7.189351e+03	8.634714e+01	5.163431e+01	-2.308108e+01	2.711149e+00
14	4.414050e+04	-1.326003e+04	1.427446e+03	1.217605e+01	1.336692e+01	-4.994748e+00	5.396399e-01
15	3.701317e+04	-1.100427e+04	1.157162e+03	1.106350e+01	1.048883e+01	-4.049508e+00	4.376645e-01
16	1.363270e+04	-3.674729e+03	3.940616e+02	2.285753e+00	4.026050e+00	-1.347994e+00	1.493497e-01
17	1.414426e+06	-1.632081e+05	7.990741e+03	1.812206e+03	3.720334e+02	-5.752442e+01	3.118402e+00
18	1.302722e+05	-1.521845e+04	5.469465e+02	8.593758e+01	4.271997e+01	-5.753978e+00	2.143897e-01
19	4.796865e+04	-5.944884e+03	2.345594e+02	2.642410e+01	1.633684e+01	-2.263892e+00	9.223716e-02
20	3.278633e+04	-4.449137e+03	2.059554e+02	4.868375e+00	1.221186e+01	-1.732838e+00	8.135019e-02
21	3.232320e+05	-4.296628e+04	1.580844e+03	8.876964e+00	1.261186e+02	-1.692144e+01	6.256486e-01
22	1.767843e+05	-1.056427e+04	6.828298e+02	8.366586e+00	7.361881e+01	-4.349568e+00	2.722543e-01
23	6.845199e+05	-1.254514e+05	1.842840e+04	9.169630e+00	2.777165e+02	-4.997925e+01	7.316040e+00
24	1.823120e+05	-2.402375e+04	9.038157e+02	-7.829785e-01	7.205974e+01	-9.501195e+00	3.580032e-01
25	5.493692e+04	-4.274782e+03	1.067321e+02	3.170664e+00	2.141401e+01	-1.681173e+00	4.221481e-02
26	4.355633e+05	-4.114607e+04	1.152095e+03	1.616167e+01	1.706801e+02	-1.625328e+01	4.565808e-01
27	2.203797e+05	-1.658472e+04	4.602671e+02	1.048765e+00	8.705734e+01	-6.573435e+00	1.826383e-01
28	1.031312e+05	-1.138180e+04	3.585158e+02	4.622451e+00	4.044276e+01	-4.504334e+00	1.422007e-01
29	2.075588e+05	-2.812610e+04	1.832678e+03	2.416417e+02	5.553028e+01	-1.015681e+01	7.168989e-01
30	3.969873e+04	-4.365093e+03	1.703038e+02	2.003748e+01	1.370233e+01	-1.668641e+00	6.702659e-02

Table C.8: Coefficients for single scattering albedo ω in Eq. 2.74 for RRTM wavelength bands.

	c_0	c_1	c_2	c_3	c_4	c_5	c_6	c_7
1	7.482969e+02	1.191845e+00	-2.512046e-01	6.347730e-03	5.895960e+02	4.777691e+01	-1.050961e+00	1.485196e-02
2	5.835067e+02	3.239986e+02	-2.136009e+01	8.792269e-01	-2.577041e+03	1.624716e+03	-3.830457e+01	1.882829e+00
3	-4.704665e+00	9.978438e+00	-8.897123e-01	3.273509e-02	-5.851676e+01	3.205800e+01	-2.249147e+00	7.333297e-02
4	6.092744e+01	2.227758e+02	-2.189104e+01	1.024904e+00	-9.871535e+02	6.938920e+02	-5.761561e+01	2.316505e+00
5	3.185261e+02	1.194548e+02	-8.298239e+00	3.750373e-01	-1.722323e+02	3.393487e+02	-2.213338e+01	8.463917e-01
6	9.256759e+03	9.952278e+02	-1.051704e+02	5.301405e+00	4.485896e+03	3.372461e+03	-2.620872e+02	1.137447e+01
7	1.932220e+01	2.729757e+01	-1.923483e+00	7.721231e-02	-1.317817e+02	9.385588e+01	-4.529994e+00	1.638937e-01
8	-7.558231e+01	4.243644e+01	-3.628684e+00	1.556277e-01	-3.098579e+02	1.327022e+02	-8.178000e+00	3.323719e-01
9	-1.288537e+01	1.750942e+01	-1.761674e+00	1.257391e-01	-1.208086e+02	6.979040e+01	-4.229560e+00	2.704516e-01
10	-4.162814e+01	2.017393e+01	-2.523227e+00	1.424081e-01	-1.245275e+02	5.390629e+01	-5.527350e+00	3.052596e-01
11	-5.431678e+01	2.431395e+01	-3.159876e+00	1.765723e-01	-1.454011e+02	6.020461e+01	-6.837211e+00	3.781122e-01
12	1.510932e+02	-5.555472e+01	9.360776e+00	2.737921e-02	-1.667890e+02	7.698727e+01	1.880557e+01	6.073608e-02
13	2.233811e+02	-8.491849e+01	1.211484e+01	8.992548e-02	1.274606e+02	-1.370624e+00	2.472151e+01	1.959970e-01
14	1.214592e+02	-4.577831e+01	6.326319e+00	9.142501e-02	1.311621e+02	-1.455673e+01	1.282646e+01	1.987612e-01
15	3.690207e+01	-2.072192e+01	3.271244e+00	2.497560e-02	-1.686343e+02	3.455161e+01	6.485055e+00	5.492962e-02
16	2.455326e+01	-1.573754e+01	2.516702e+00	2.143335e-02	-2.441568e+02	5.128985e+01	4.838573e+00	4.734589e-02
17	2.144097e+02	-4.293264e+01	5.164097e+00	6.036715e-02	8.621745e+02	-1.283853e+02	2.095509e+01	1.156995e-01
18	2.996771e+00	7.944797e-01	1.455449e-02	1.296372e-04	7.334823e+02	6.489375e+00	1.584060e-01	1.803706e-04
19	8.111848e+00	3.158661e+00	-3.470314e-02	8.522681e-04	7.052618e+02	-3.469081e+00	1.143583e-01	1.596854e-03
20	6.767090e-02	1.686542e+00	3.692069e-02	1.179050e-04	1.557317e+03	3.004536e+01	2.431321e-01	1.381278e-04
21	-7.916927e-02	5.590768e-01	2.455904e-03	3.101830e-05	4.748530e+02	2.645087e+00	4.691439e-02	7.096310e-05
22	7.107928e-03	8.739532e-03	-6.087625e-05	1.139308e-06	9.751998e+01	-7.151822e-01	1.238603e-02	1.620324e-06
23	1.412611e-04	7.509057e-04	1.730238e-05	2.101188e-08	5.745138e+01	1.251058e+00	1.795473e-03	-6.425294e-08
24	3.540041e-07	2.361418e-06	-7.226781e-07	5.459253e-08	5.445892e+00	-1.625223e+00	1.224099e-01	-4.951246e-07
25	-3.502750e-07	1.767480e-07	-2.761199e-08	1.361639e-09	2.337912e+00	-4.761993e-01	2.447114e-02	3.912835e-07
26	-1.328948e-07	7.131830e-08	-1.187949e-08	6.276858e-10	6.968111e-01	-1.590158e-01	9.128262e-03	2.294752e-07
27	-4.210711e-07	4.083336e-07	-3.365474e-08	2.056720e-09	2.014986e+00	-1.882357e-01	1.252706e-02	1.979652e-07
28	-6.973571e-06	1.520221e-06	-1.151295e-06	1.850199e-07	2.239693e-01	-2.759900e+00	4.664720e-01	-2.365198e-06
29	3.634076e+01	-8.467392e+00	9.100310e-01	1.616390e-03	9.696409e+01	-9.904289e+00	1.896428e+00	3.536500e-03
30	7.622126e+01	2.858602e+00	2.516827e-01	1.809691e-04	2.483189e+02	7.154815e+00	5.507016e-01	4.023678e-04

Table C.9: Coefficients for asymmetry parameter g in Eq. 2.75 for RRTM wavelength bands.

	c_0	c_1	c_2	c_3	c_4	c_5	c_6	c_7	c_8
1	3.524690e+00	-1.664196e+01	2.823449e+01	-1.468670e+01	3.707969e+00	-1.729027e+01	2.952906e+01	-1.533064e+01	3.107914e-01
2	9.603133e-01	-3.827071e+00	5.221643e+00	-2.353083e+00	9.602504e-01	-3.611309e+00	4.372098e+00	-1.177431e+00	-5.411807e-01
3	1.520245e+00	-4.763848e+00	4.843756e+00	-1.572739e+00	1.606352e+00	-5.030005e+00	5.186062e+00	-1.847349e+00	1.194165e-01
4	1.320785e+00	-4.373261e+00	4.785480e+00	-1.723814e+00	1.407364e+00	-4.640318e+00	5.052940e+00	-1.812353e+00	3.527659e-03
5	1.202055e+00	-4.053002e+00	4.524655e+00	-1.668856e+00	1.285669e+00	-4.314229e+00	4.775607e+00	-1.721897e+00	-1.952038e-02
6	-5.306365e-01	2.971334e+00	-2.600249e+00	2.652645e-01	-5.499616e-01	3.077766e+00	-2.695992e+00	3.254734e-01	-4.100449e-02
7	8.404088e-01	-2.499731e+00	2.312251e+00	-6.251807e-01	8.549724e-01	-2.537736e+00	2.362997e+00	-6.876572e-01	3.752694e-02
8	2.218431e+00	-6.117416e+00	4.963068e+00	-9.600735e-01	2.282105e+00	-6.351195e+00	5.426926e+00	-1.519998e+00	2.777278e-01
9	1.189001e+00	-3.398189e+00	3.014695e+00	-7.635532e-01	1.231427e+00	-3.563508e+00	3.328393e+00	-1.100772e+00	1.520294e-01
10	9.379358e-01	-3.058556e+00	3.267874e+00	-1.133863e+00	9.695696e-01	-3.172663e+00	3.430862e+00	-1.252953e+00	4.006659e-02
11	9.568631e-01	-3.165757e+00	3.439754e+00	-1.219654e+00	9.874925e-01	-3.273621e+00	3.583914e+00	-1.313352e+00	2.790615e-02
12	1.006626e+00	-3.217981e+00	3.322819e+00	-1.087053e+00	1.038377e+00	-3.325466e+00	3.497581e+00	-1.269900e+00	8.839307e-02
13	8.093362e-01	-2.667734e+00	2.881652e+00	-1.012972e+00	8.381489e-01	-2.769298e+00	3.027596e+00	-1.126673e+00	4.224597e-02
14	7.509604e-01	-2.489641e+00	2.715437e+00	-9.697943e-01	7.795417e-01	-2.593986e+00	2.865385e+00	-1.078172e+00	3.528733e-02
15	8.159209e-01	-2.683585e+00	2.885353e+00	-1.005564e+00	8.471120e-01	-2.794816e+00	3.054773e+00	-1.151829e+00	5.938623e-02
16	2.936628e+00	-9.616839e+00	1.026625e+01	-3.534883e+00	3.049690e+00	-1.001782e+01	1.090289e+01	-4.134256e+00	2.624788e-01
17	-8.405115e-01	4.745654e+00	-4.064599e+00	2.828752e-01	-8.622531e-01	4.800773e+00	-3.797191e+00	0	0
18	2.590873e+00	-2.539930e+01	6.384803e+01	-8.753067e+00	2.759321e+00	-2.660910e+01	6.394117e+01	0	0
19	-9.282617e-01	5.265259e+00	-4.738268e+00	6.452199e-01	-9.448973e-01	5.200372e+00	-3.952076e+00	0	0
20	-8.767301e-01	-3.075562e+00	3.336336e+01	-4.363404e+00	-8.552338e-01	-4.210774e+00	3.714391e+01	0	0
21	-8.370564e-01	3.502372e+00	1.529057e+00	-1.171106e-01	-8.896113e-01	3.492658e+00	2.646594e+00	0	0
22	-7.216590e-01	3.509878e+00	-1.077200e+00	2.494193e-01	-7.940595e-01	3.692840e+00	-3.549937e-01	0	0
23	2.686948e+00	-3.053494e+00	4.035289e+00	-1.638484e-01	3.042223e+00	-3.194032e+00	4.706923e+00	0	0
24	1.091724e+00	1.044387e+00	6.231003e-01	2.947317e-01	1.263527e+00	1.088340e+00	1.622535e+00	0	0
25	8.140801e-01	2.775859e+00	-1.360369e+00	5.670451e-01	9.529834e-01	3.006853e+00	-3.108990e-01	0	0
26	1.254410e+00	4.565848e+00	-2.950619e+00	9.400567e-01	1.467056e+00	5.046819e+00	-1.510103e+00	0	0
27	1.066567e+00	1.318597e+00	-1.118476e+00	3.962435e-01	1.235592e+00	1.564489e+00	-5.934078e-01	0	0
28	1.285084e+01	-2.582937e+01	1.541817e+01	-2.121008e+00	1.485234e+01	-2.705276e+01	1.263195e+01	0	0
29	4.013696e+00	-6.027151e+00	2.250685e+00	-1.740339e-02	4.138480e+00	-6.175372e+00	2.301104e+00	0	0
30	-1.333911e-01	8.245062e-01	-5.806557e-01	-1.011455e-02	-1.413399e-01	8.741809e-01	-6.163912e-01	0	0

Table C.10: Coefficients for extinction coefficient β_{ext} in Eq. 2.77 for RRTM wavelength bands.

	$c_{1,0}$	$c_{1,1}$	$c_{1,2}$	$c_{2,0}$	$c_{2,1}$	$c_{2,2}$	$c_{3,0}$	$c_{3,1}$
1	-5.07099001e+01	-9.17237976e-01	8.11504466e+01	1.16256915e+05	-5.04128978e-05	-1.16166868e+05	3.00063283e+03	-1.12790502e+00
2	-5.55784539e-02	2.89421894e+00	2.28892362e+02	1.78210837e+03	-9.27993404e-01	-1.51372059e+01	2.69571057e+03	-1.11481061e+00
3	1.08876269e+06	-1.34655511e-04	-1.08824502e+06	1.80379697e+03	-9.71573640e-01	-7.90732594e+00	2.40762178e+03	-1.09286233e+00
4	1.05352772e+06	-1.32842254e-04	-1.05303876e+06	1.47476401e+03	-9.10099138e-01	-1.02430485e+01	2.18847200e+03	-1.07382784e+00
5	7.56572188e+05	-1.12401745e-04	-7.56234894e+05	8.05754479e+02	-6.51041397e-01	-3.38046864e+01	1.91757633e+03	-1.04606772e+00
6	-4.64436148e-02	2.62390133e+00	1.52372088e+02	6.03647094e+02	-2.63502577e-01	-1.90716560e+02	2.22936896e+03	-1.08239274e+00
7	-8.57745114e-06	6.11699520e+00	2.02192211e+02	3.93576646e+03	-1.26566305e+00	4.02521532e+00	2.47653789e+03	-1.10633475e+00
8	-6.55714401e-04	4.73963377e+00	2.57334764e+02	8.81221580e+03	-1.61350538e+00	2.02091013e+01	2.36749005e+03	-1.09574426e+00
9	-7.70434426e-02	3.07016822e+00	3.17169626e+02	6.33224187e+03	-1.50724848e+00	1.93862161e+01	2.30696147e+03	-1.09044089e+00
10	-3.21406367e+01	1.01693525e+00	5.49689786e+02	3.57599318e+03	-1.28958959e+00	1.16559470e+01	2.24747840e+03	-1.08489616e+00
11	1.63014718e+06	-1.65205687e-04	-1.62932923e+06	2.79337846e+03	-1.19098190e+00	6.97159801e+00	2.17303259e+03	-1.07862963e+00
12	2.03638014e+03	-3.72488138e-01	-6.71551261e+02	2.31769043e+03	-1.11541143e+00	2.56538833e+00	2.13801873e+03	-1.07631837e+00
13	2.89875012e+03	-1.06121966e+00	-6.96488961e+01	2.27988840e+03	-1.11015269e+00	2.72924351e+00	2.09458517e+03	-1.07116261e+00
14	4.16014732e+03	-1.39531224e+00	1.13927690e+01	2.27539879e+03	-1.11716021e+00	3.10940540e+00	2.07049432e+03	-1.06924807e+00
15	5.45328815e+03	-1.65510253e+00	5.41038325e+01	2.25325457e+03	-1.11668372e+00	3.29663043e+00	2.04858878e+03	-1.06737919e+00
16	4.10147364e+03	-1.55585936e+00	5.77792494e+01	2.10782776e+03	-1.09724313e+00	2.88570599e+00	2.00039460e+03	-1.06404339e+00
17	3.14495568e+03	-1.37667534e+00	3.81063771e+01	2.05941330e+03	-1.08817071e+00	2.54984454e+00	1.97024406e+03	-1.06106841e+00
18	3.63432626e+03	-1.46674766e+00	4.51734594e+01	2.04029400e+03	-1.08727641e+00	2.56408363e+00	1.93848974e+03	-1.05743384e+00
19	3.21486760e+03	-1.43921852e+00	5.06655534e+01	1.99668881e+03	-1.08124654e+00	2.52577306e+00	1.90862807e+03	-1.05517294e+00
20	2.54167169e+03	-1.27972742e+00	3.32987877e+01	1.94311644e+03	-1.07403837e+00	2.19732881e+00	1.88706097e+03	-1.05247065e+00
21	2.09196067e+03	-1.13458712e+00	1.23204339e+01	1.89594892e+03	-1.06630330e+00	1.96434681e+00	1.86492758e+03	-1.04998765e+00
22	1.98177352e+03	-1.10892372e+00	8.07609310e+00	1.83431092e+03	-1.05805211e+00	1.67610937e+00	1.83633715e+03	-1.04633961e+00
23	1.96433717e+03	-1.10991214e+00	8.89337024e+00	1.80686051e+03	-1.05230709e+00	1.53261627e+00	1.82220713e+03	-1.04459915e+00
24	1.88068451e+03	-1.09300518e+00	8.16607962e+00	1.74945678e+03	-1.04371566e+00	1.24779585e+00	1.79063195e+03	-1.04209360e+00
25	1.80299189e+03	-1.07606838e+00	6.70573328e+00	1.69597928e+03	-1.03535956e+00	1.02179333e+00	1.76132921e+03	-1.03899846e+00
26	1.73726926e+03	-1.06185583e+00	5.25451377e+00	1.66407118e+03	-1.02938852e+00	8.25872629e-01	1.74202663e+03	-1.03679727e+00
27	1.68738634e+03	-1.04867516e+00	4.23564982e+00	1.63588131e+03	-1.02427773e+00	7.27885638e-01	1.72761073e+03	-1.03553937e+00
28	1.63442450e+03	-1.02936419e+00	1.43032117e+00	1.62169095e+03	-1.02488423e+00	8.08507486e-01	1.71460060e+03	-1.03394889e+00
29	1.53323759e+03	-1.00077530e+00	-6.03778846e+00	1.66354595e+03	-1.04658025e+00	1.71704366e+00	1.71002667e+03	-1.03389625e+00
30	-8.90965349e-01	1.89852664e+00	2.38139220e+02	1.70689832e+03	-9.72358151e-01	-7.93626007e+00	2.12924367e+03	-1.07480228e+00

Table C.11: Coefficients for asymmetry parameter g in Eq. 2.77 for RRTM wavelength bands.

	$c_{1,0}$	$c_{1,1}$	$c_{1,2}$	$c_{2,0}$	$c_{2,1}$	$c_{2,2}$	$c_{3,0}$	$c_{3,1}$	$c_{3,2}$
1	-4.25116586e+00	-1.16827084e-01	3.82707882e+00	-4.36134267e+00	-1.08486211e+00	9.36361359e-01	-7.79287508e+00	-1.28523392e+00	9.26540799e-01
2	-2.33174602e+00	-9.89468461e-01	1.00604221e+00	-3.23310705e+00	-1.26934508e+00	9.41404392e-01	-9.83049595e+00	-1.67440631e+00	9.30503979e-01
3	-2.51019515e+00	-1.26300851e+00	9.56190412e-01	-3.13513208e+00	-1.42976239e+00	9.36884872e-01	-4.44960986e+00	-1.55586808e+00	9.34874100e-01
4	-2.52995212e+00	-1.41544607e+00	9.52374263e-01	-2.63099872e+00	-1.46242191e+00	9.46112103e-01	-2.91344123e+00	-1.49128932e+00	9.46112479e-01
5	-2.35941361e+00	-1.47808835e+00	9.67130961e-01	-2.26130651e+00	-1.47472917e+00	9.64942164e-01	-2.27407895e+00	-1.46675759e+00	9.65362409e-01
6	-1.99764800e+00	-1.52536128e+00	9.78481640e-01	-9.16103984e-01	-1.20168303e+00	9.80651161e-01	-8.60024569e-01	-1.04613339e+00	9.89538553e-01
7	-2.04684114e+00	-1.95214475e+00	9.14962846e-01	1.78779124e-03	1.09970188e+00	8.67900244e-01	-3.03328651e+01	-1.96678419e+00	9.79114581e-01
8	-2.68799901e+00	-2.29273598e+00	9.10493814e-01	4.99489348e-05	2.05926210e+00	8.86752865e-01	-7.43598848e+01	-2.21435660e+00	9.79077613e-01
9	-6.07256968e+00	-3.25453304e+00	8.86007580e-01	1.82371554e-02	6.47307223e-01	7.79443507e-01	-8.47401730e+01	-2.29489727e+00	9.75685328e-01
10	-3.14135648e+01	-4.82005384e+00	8.69381052e-01	-1.09938212e+00	-6.82801838e-01	1.05400456e+00	-7.58410865e+01	-2.36911251e+00	9.72188863e-01
11	-3.68358225e-01	-1.35249289e+00	9.13048645e-01	-2.62315822e+00	-1.35856130e+00	9.83738797e-01	-1.79557549e+01	-2.11250075e+00	9.71774199e-01
12	-1.51862723e+00	1.44850101e-02	2.40257247e+00	-1.72610634e+00	-9.26749272e-01	1.00191123e+00	-3.10272639e+00	-1.17321181e+00	9.85276589e-01
13	7.27472064e-15	1.10796415e+01	8.26348641e-01	-2.18341445e+00	-1.10509705e+00	9.83827845e-01	-4.14120536e+00	-1.31406913e+00	9.80577420e-01
14	8.62880044e-09	6.02010734e+00	8.07930833e-01	-1.37603608e+00	-9.14032859e-01	9.81476229e-01	-1.69333348e+00	-9.46309420e-01	9.87310151e-01
15	2.82281635e-08	5.71036851e+00	7.85998698e-01	-7.79527093e-01	-6.42599504e-01	9.87396408e-01	-7.85667278e-01	-5.77060667e-01	1.00928469e+00
16	1.11780039e-03	1.72548806e+00	8.04054243e-01	-4.45322334e-01	-6.09650787e-01	9.80933164e-01	-4.99278672e-01	-6.29829105e-01	9.82880186e-01
17	-5.42581960e+03	-9.53305334e-06	5.42661951e+03	-4.71396734e-01	-1.06776808e+00	9.60183075e-01	-1.87079937e-01	-5.74263106e-01	9.73816891e-01
18	-3.84361455e+03	-8.01435285e-06	3.84444939e+03	-5.16543728e-01	-9.96499880e-01	9.57435574e-01	-1.95864764e-01	-4.31825599e-01	9.85421469e-01
19	2.51059564e-04	2.13309560e+00	8.19948508e-01	-4.05057564e-01	-7.02716776e-01	9.38760925e-01	-2.63900372e-01	-3.37664593e-01	9.87846123e-01
20	-6.22487809e+03	-1.02106971e-05	6.22557407e+03	-2.75269267e-01	-4.62927198e-01	9.39886953e-01	-2.56322990e-01	-3.31377055e-01	9.67800936e-01
21	-3.56782679e-01	-7.37747068e-01	9.09106131e-01	-2.48569177e-01	-5.24062594e-01	9.18888619e-01	-2.28639864e-01	-4.39757519e-01	9.28488974e-01
22	-4.99651276e-01	-1.33484556e+00	8.70815071e-01	-2.27376461e-01	-6.16804476e-01	9.01994787e-01	-2.25610079e-01	-6.01913771e-01	9.02688901e-01
23	-3.51839054e-01	-1.26898014e+00	8.71002193e-01	-2.09281918e-01	-6.58509960e-01	8.95277221e-01	-2.65232993e-01	-7.68480416e-01	8.91917749e-01
24	-2.72334879e-01	-1.10287502e+00	8.74359311e-01	-2.03822718e-01	-7.07844913e-01	8.91146007e-01	-2.80189879e-01	-8.53025054e-01	8.88247301e-01
25	-1.62137730e-01	-8.81666508e-01	8.81435710e-01	-1.37474815e-01	-7.29130137e-01	8.86781909e-01	-2.32475150e-01	-9.49265605e-01	8.84308334e-01
26	-1.18798727e-01	-6.43097316e-01	8.90442269e-01	-1.28126475e-01	-7.73053493e-01	8.85476135e-01	-1.94158140e-01	-9.54875852e-01	8.83718363e-01
27	-7.96107270e-02	-7.21909879e-01	8.81901479e-01	-1.20387257e-01	-8.97116223e-01	8.81328309e-01	-1.97391156e-01	-1.08128930e+00	8.80444723e-01
28	-8.42843060e-02	-2.64422482e-01	9.13812976e-01	-1.97549895e-01	-1.24783060e+00	8.78717214e-01	-1.10471370e-01	-9.37997261e-01	8.80137055e-01
29	3.17562200e-04	1.53851448e+00	8.54694643e-01	-5.26191416e+00	-2.66625143e+00	8.75358808e-01	-5.80510562e-02	-7.43300659e-01	8.78398382e-01
30	-2.01008975e-01	-1.43391047e-01	1.00233401e+00	-9.33743811e-01	-7.56401310e-01	1.00460562e+00	-2.61602467e+00	-1.20169339e+00	9.76954878e-01

Table C.12: Coefficients for single scattering albedo ω in Eq. 2.77 for RRTM wavelength bands.

	$c_{1,0}$	$c_{1,1}$	$c_{1,2}$	$c_{2,0}$	$c_{2,1}$	$c_{2,2}$	$c_{3,0}$	$c_{3,1}$	$c_{3,2}$
1	1.13479236e+00	-5.98565179e-01	3.21296065e-01	1.55279414e+01	-2.24194293e+00	5.20746355e-01	-5.48739578e-03	6.24344596e-01	5.74037308e-01
2	1.26897092e+00	-1.19644076e+00	4.74318771e-01	3.19764520e+03	-8.36126696e-06	-3.19704083e+03	-3.67810858e-01	8.28056474e-02	1
3	9.83801351e-01	-1.01784421e+00	4.74293000e-01	4.82683830e-01	-1.71414918e-01	2.34885141e-01	-3.80484891e-01	7.85880707e-02	1
4	8.42731761e-01	-7.36646065e-01	4.26132162e-01	6.23135023e-01	-6.27897411e-01	4.30285761e-01	-3.86758206e-01	7.41737975e-02	1
5	8.64013441e-01	-4.93573987e-01	3.25379816e-01	1.32482570e+00	-9.95856523e-01	4.68233560e-01	-3.85491354e-01	6.97812803e-02	1
6	9.90316118e-01	-6.86462524e-01	2.90847917e-01	7.13650323e-07	3.12727741e+00	4.61997784e-01	-10	-1.95053645e+00	5.06984424e-01
7	1.18251550e-12	9.63818533e+00	2.72316088e-01	-3.45533952e+00	-9.81688713e-01	5.90566106e-01	-10	-1.68450010e+00	5.01109466e-01
8	2.18395578e-09	6.58919425e+00	2.36435582e-01	-9.39464777e+00	-1.40423297e+00	5.42072417e-01	-10	-1.66309657e+00	4.97706637e-01
9	1.06694355e-05	3.68739312e+00	1.92109648e-01	-1.84721030e+01	-1.77829466e+00	5.00975988e-01	-10	-1.66912985e+00	4.92520358e-01
10	6.26513047e-04	2.27427604e+00	1.67991574e-01	-1.95022165e+01	-1.95290135e+00	4.90990431e-01	-10	-1.84889399e+00	4.85409117e-01
11	3.27650566e-03	1.65051817e+00	2.21650212e-01	-2.53981039e+01	-2.36558180e+00	4.85437105e-01	4.73535296e-01	2.63214829e-03	0
12	8.43569128e-03	1.29627099e+00	2.47750863e-02	-1.34913643e+00	-2.12289361e-01	1.02926453e+00	-5.64079085e+00	-1.09663098e+00	5.07660868e-01
13	8.91171091e-02	5.89368924e-01	-1.23752141e-01	-1.22133718e+00	-2.57698404e-01	8.99259663e-01	-9.68860621e+00	-1.33881336e+00	4.91863417e-01
14	1.57304815e-01	4.02483818e-01	-2.15636173e-01	-9.92201519e+03	-1.47350258e-05	9.92185794e+03	-3.11255733e+00	-8.25833061e-01	5.28320180e-01
15	1.73279570e-01	3.11744854e-01	-2.25987337e-01	-9.25440073e+03	-1.42202897e-05	9.25421727e+03	-1.49582793e+00	-3.29318055e-01	7.55858941e-01
16	-5.70653368e-01	-6.56397413e-01	3.80832049e-01	-7.04720853e+03	-1.24154345e-05	7.04726098e+03	-9.46107240e-01	-4.33099242e-01	5.67048272e-01
17	-9.24651373e-01	-1.40283729e+00	4.08938256e-01	-3.65176069e-01	-1.68259228e-01	6.23103845e-01	-3.73606120e-01	-4.20591395e-01	5.05615765e-01
18	-5.04264715e-01	-3.87342516e-01	4.37429255e-01	-5.63258371e+03	-1.10998750e-05	5.63267320e+03	-1.13993996e+00	-8.06268610e-02	1.16754474e+00
19	5.65824841e-02	3.40390583e-01	-6.62413448e-02	1.52051155e-02	6.10923236e-01	-3.81396077e-03	9.25679184e-02	3.05390573e-01	-1.44236341e-01
20	7.18771685e-03	6.50546820e-01	-8.77677290e-03	3.71965776e-03	8.29513968e-01	-1.81441278e-03	2.75165527e-02	4.45089413e-01	-6.46402546e-02
21	3.00980576e-03	7.14204487e-01	-3.25133000e-03	1.60764638e-03	8.84241186e-01	6.86703887e-06	9.55690424e-03	5.32267126e-01	-2.59380511e-02
22	1.05007742e-03	7.55023889e-01	-9.94900791e-04	5.24691788e-04	9.48409866e-01	3.10960282e-04	2.39494022e-03	6.38922853e-01	-7.67665471e-03
23	4.63373757e-04	7.77425139e-01	-3.83048615e-04	2.80143419e-04	9.25974098e-01	1.01954943e-05	9.86582167e-04	6.73933399e-01	-3.26763751e-03
24	6.87556242e-05	7.64639625e-01	-5.39264448e-05	3.98136395e-05	9.28853638e-01	7.37924702e-06	1.39134102e-04	6.79943754e-01	-4.65191867e-04
25	5.28950568e-07	1.04045938e+00	3.80423353e-07	7.43340337e-07	9.36179720e-01	-2.26565284e-08	4.45159188e-06	5.76227365e-01	-1.37080248e-05
26	5.29975226e-08	9.79683148e-01	2.56333274e-08	7.95969986e-08	9.13212515e-01	-5.69009340e-08	1.84812272e-07	7.14991795e-01	-2.78167558e-07
27	-1.00000289e+00	-7.43621272e-07	1.00000236e+00	2.52442247e-07	7.80874830e-01	-4.43160015e-07	2.93217750e-07	7.73273017e-01	-9.05167165e-07
28	-2.14457652e+00	-1.74480016e-06	2.14457374e+00	6.48042713e-07	9.50142526e-01	-2.85039203e-07	2.20439825e-06	7.02314770e-01	-7.97989859e-06
29	3.50626277e-07	1.37044681e+00	1.21798208e-06	1.53782107e-06	9.09682459e-01	-3.42840773e-06	4.64775107e-06	6.86334084e-01	-1.78141529e-05
30	5.85274855e-02	6.53902149e-01	-3.74575707e-02	-1.12629632e+00	-4.77105822e-01	6.00694707e-01	-3.00689260e+00	-9.46502925e-01	4.98652622e-01

D Sensitivity of seeding to background conditions

Table D.1: Changes in LW radiative fluxes averaged over entire domain and simulation time and assumptions made for the sensitivity to background conditions. All fluxes in W m^{-2}.

	Δ net, TOA	Δ net, SFC	remarks
TKE0	-6.45 ± 0.27	-4.11 ± 0.38	$f_w = 0.3$
TKE1	-7.17 ± 0.34	-4.10 ± 0.39	$f_w = 0.2$
TKE2	-4.89 ± 0.17	-3.28 ± 0.25	$f_w = 0.4$
HOM0	-6.45 ± 0.27	-4.11 ± 0.38	$n_{HOM} = 1000$ cm^{-1}; $d_{HOM} = 200$ nm
HOM1	-7.00 ± 0.31	-4.23 ± 0.39	$n_{HOM} = 200$ cm^{-1}; $d_{HOM} = 40$ nm
HOM2	-7.06 ± 0.31	-4.24 ± 0.39	$n_{HOM} = 200$ cm^{-1}; $d_{HOM} = 20$ nm
HET0	-6.90 ± 0.29	-6.95 ± 0.65	$n_{DUST} = 0$
HET1	-6.45 ± 0.27	-4.11 ± 0.38	-
HET2	-6.48 ± 0.28	-0.81 ± 0.06	$10 \times n_{DUST}$

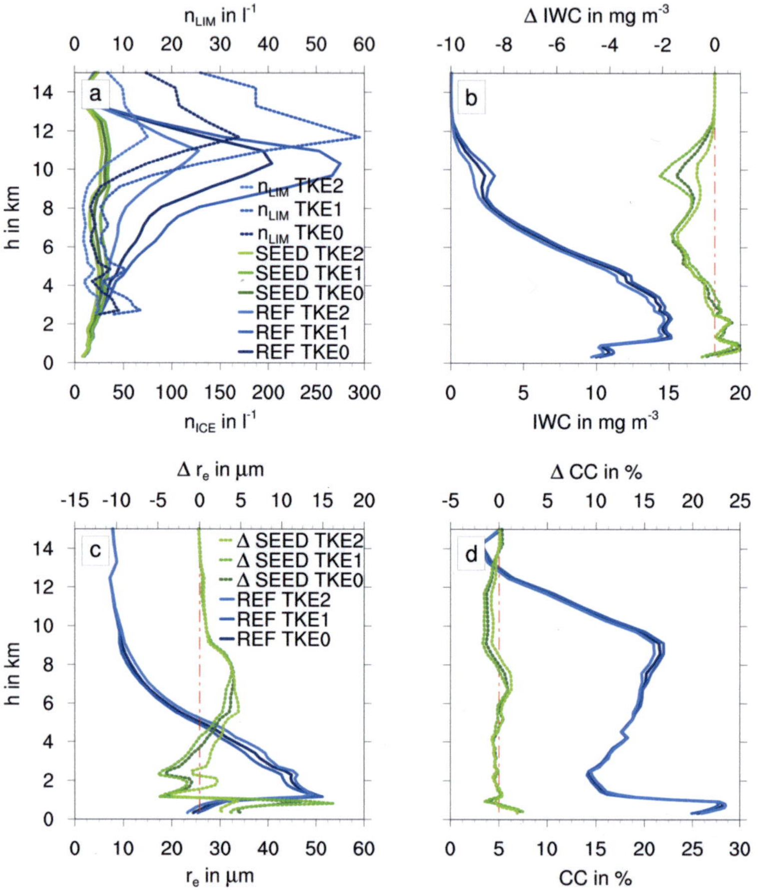

Figure D.1: Comparison of ice cloud properties for different values of f_{w} for averaged over the entire domain and simulation time for reference simulations (blue) and simulations with $n_{\mathrm{SEED}} = 10^3$ l^{-1} and $d_{\mathrm{SEED}} = 6.7$ μm. a) n_{ICE}; b) IWC; c) r_{e}; d) cloud cover. Solid lines are absolute values, dotted lines represent differences compared to the corresponding reference simulation. Note that in a) only absolute values are depicted, the dotted lines are n_{LIM} for the reference simulations.

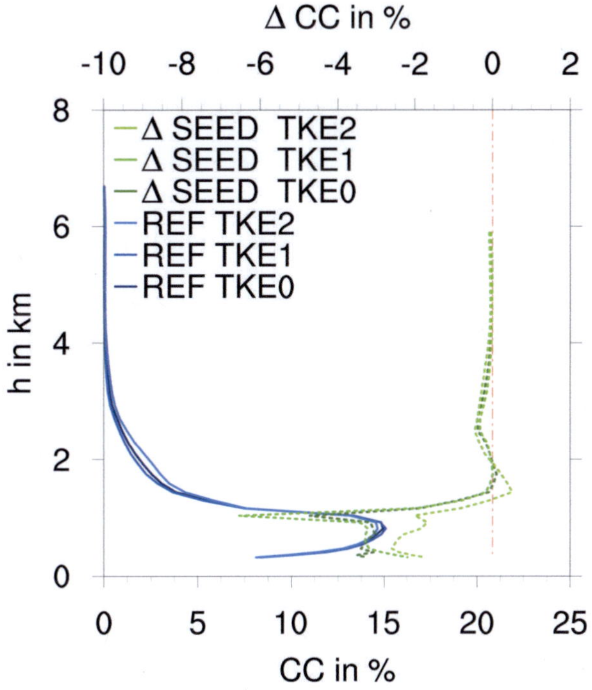

Figure D.2: Comparison of liquid (mixed-phase) cloud cover for different values of f_w averaged over the entire domain and simulation time for reference simulations (blue) and simulations with $n_{SEED} = 10^3$ l^{-1} and $d_{SEED} = 6.7\ \mu$m. Solid lines are absolute values, dotted lines represent differences compared to the reference simulation.

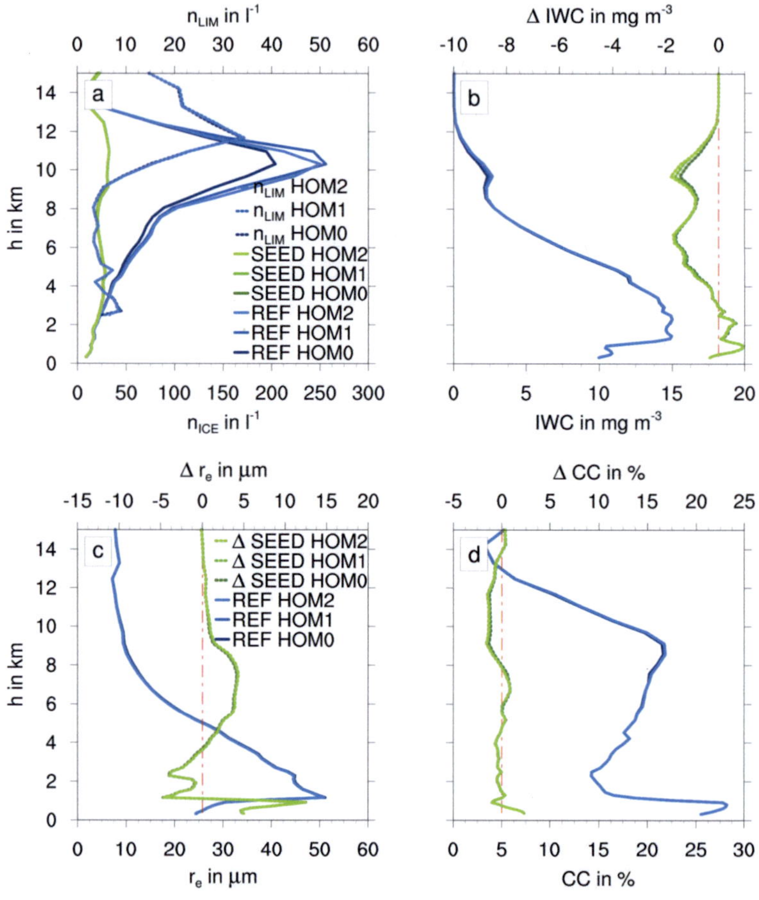

Figure D.3: Comparison of ice cloud properties for different sulphate aerosol particle concentrations and sizes averaged over the entire domain and simulation time for reference simulations (blue) and simulations with $n_{SEED} = 10^3$ l^{-1} and $d_{SEED} = 6.7$ μm. a) n_{ICE}; b) IWC; c) r_e; d) cloud cover. Solid lines are absolute values, dotted lines represent differences compared to the corresponding reference simulation. Note that in a) only absolute values are depicted, the dotted lines are n_{LIM} for the reference simulations.

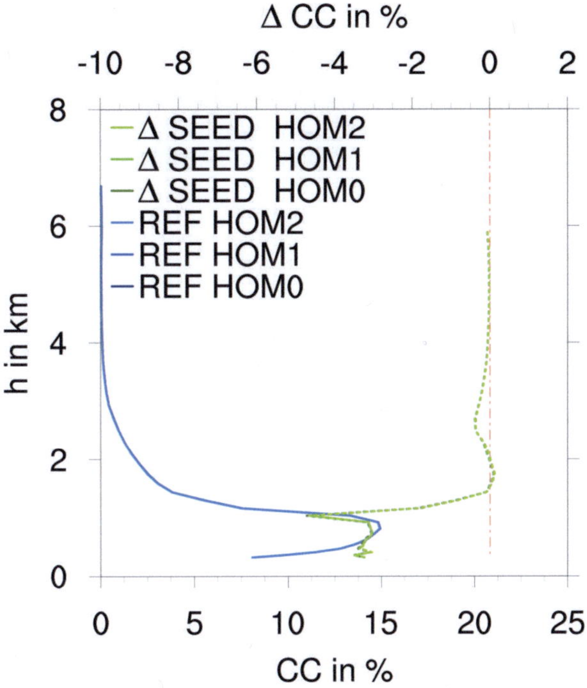

Figure D.4: Comparison of liquid (mixed-phase) cloud cover for different sulphate aerosol particle concentrations and sizes averaged over the entire domain and simulation time for reference simulations (blue) and simulations with $n_{\mathrm{SEED}} = 10^3\ \mathrm{l}^{-1}$ and $d_{\mathrm{SEED}} = 6.7\ \mu\mathrm{m}$. Solid lines are absolute values, dotted lines represent differences compared to the reference simulation.

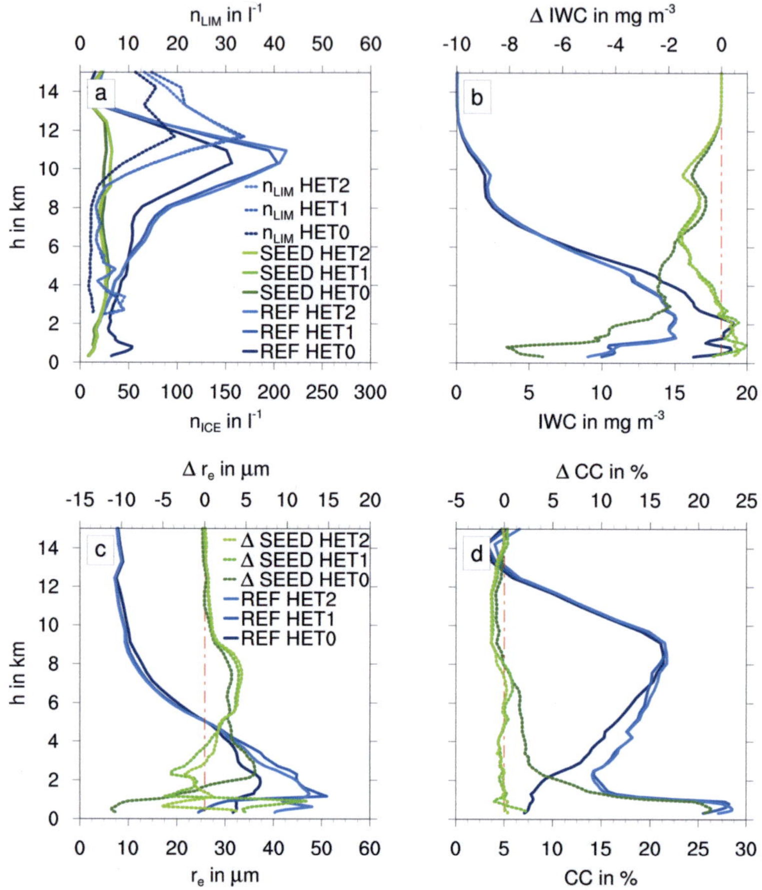

Figure D.5: Comparison of ice cloud properties for different dust aerosol concentrations averaged over the entire domain and simulation time for reference simulations (blue) and simulations with $n_{\mathrm{SEED}} = 10^3$ l^{-1} and $d_{\mathrm{SEED}} = 6.7$ μm. a) n_{ICE}; b) IWC; c) r_{e}; d) cloud cover. Solid lines are absolute values, dotted lines represent differences compared to the corresponding reference simulation. Note that in a) only absolute values are depicted, the dotted lines are n_{LIM} for the reference simulations.

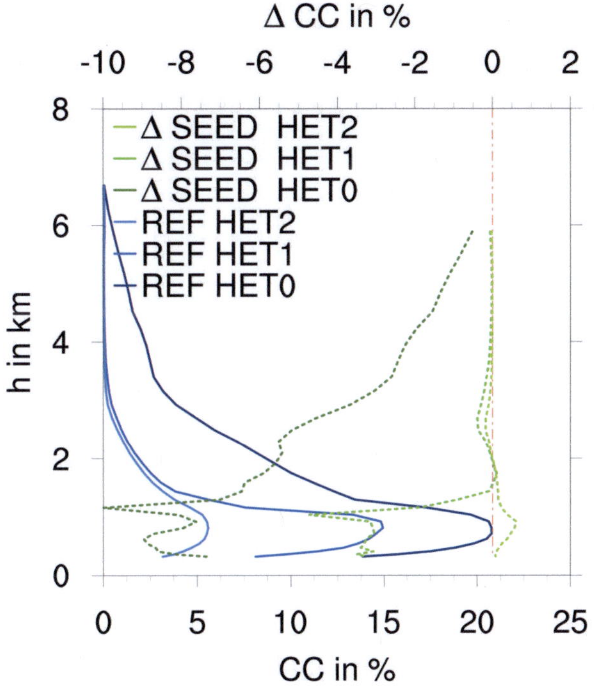

Figure D.6: Comparison of liquid (mixed-phase) cloud cover for different dust aerosol concentrations averaged over the entire domain and simulation time for reference simulations (blue) and simulations with $n_{\mathrm{SEED}} = 10^3$ l^{-1} and $d_{\mathrm{SEED}} = 6.7$ μm. Solid lines are absolute values, dotted lines represent differences compared to the reference simulation.

E Comparison to GLORIA data

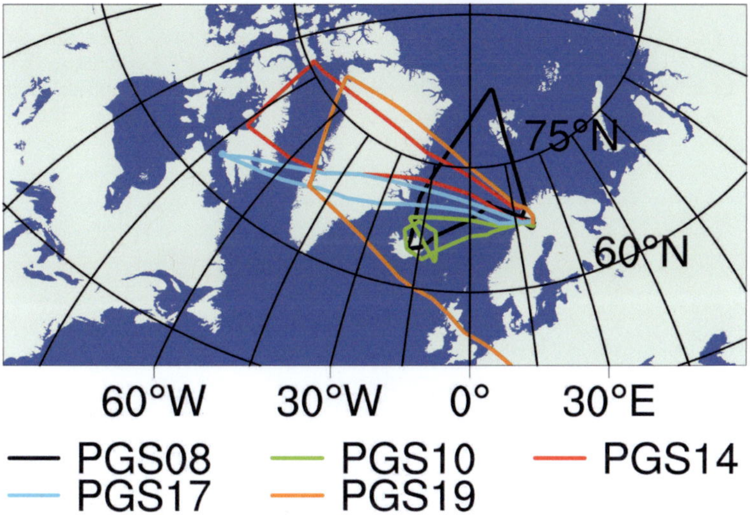

Figure E.1: Flight paths for all HALO flights considered.

PGS10, 25 January 2016

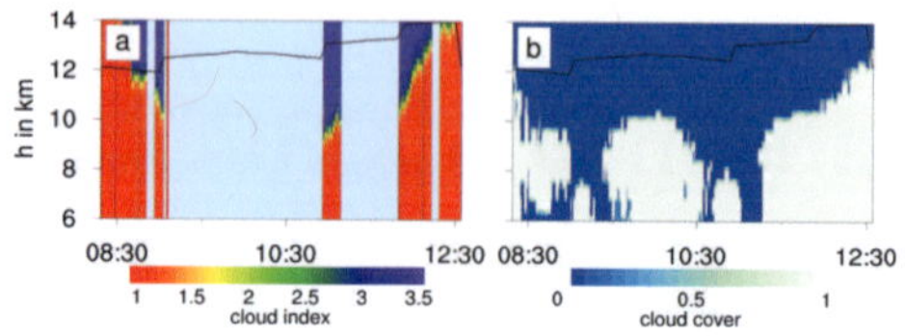

Figure E.2: (a) Vertical cross-section of cloud index derived from GLORIA measurements during PGS10 on 25 January 2016. Low cloud index values indicate opaque conditions due to cloud particles along the instrument's line of sight. (b) Cloud mask derived from the model simulation *REF07* at the geolocations of the measurement (0 = no cloud / 1 = cloud present). Black line in both panels: HALO flight altitude. Times in UTC.

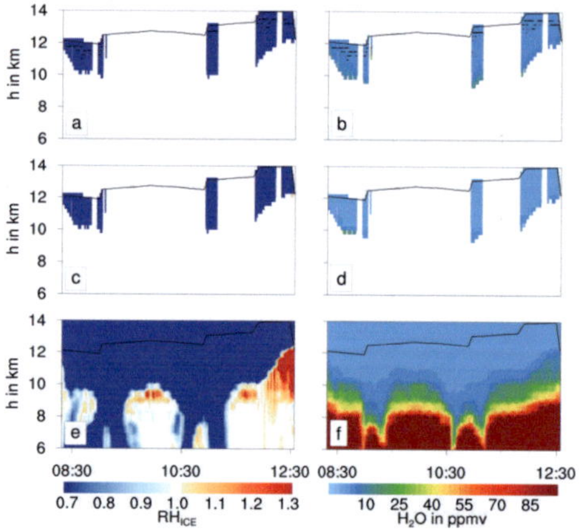

Figure E.3: Cross sections of relative humidity (i.e. supersaturation versus ice phase) (a) and water vapor volume mixing ratio (b) derived from GLORIA measurements during the HALO flight PGS10 on 25 January 2016. Relative humidity (c) and water vapor volume mixing ratio (d) simulated by the model and filtered for geolocations where GLORIA data are available. Full vertical cross section of the same parameters from the model simulation *REF07* (e,f). Times in UTC.

PGS14, 26 February 2016

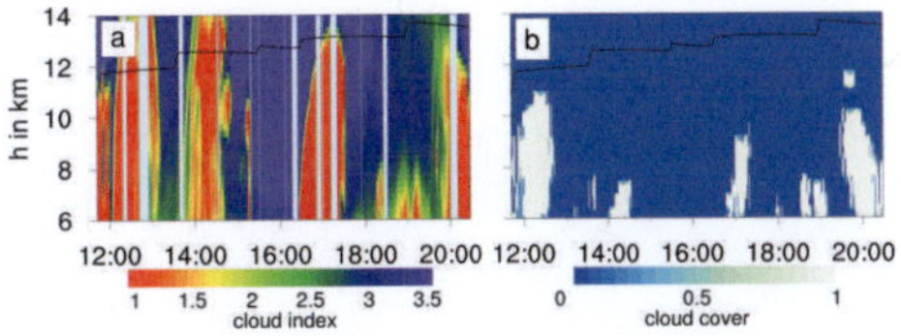

Figure E.4: (a) Vertical cross-section of cloud index derived from GLORIA measurements during PGS14 on 26 February 2016. Low cloud index values indicate opaque conditions due to cloud particles along the instrument's line of sight. (b) Cloud mask derived from the model simulation *REF08* at the geolocations of the measurement (0 = no cloud / 1 = cloud present). Black line in both panels: HALO flight altitude. Times in UTC.

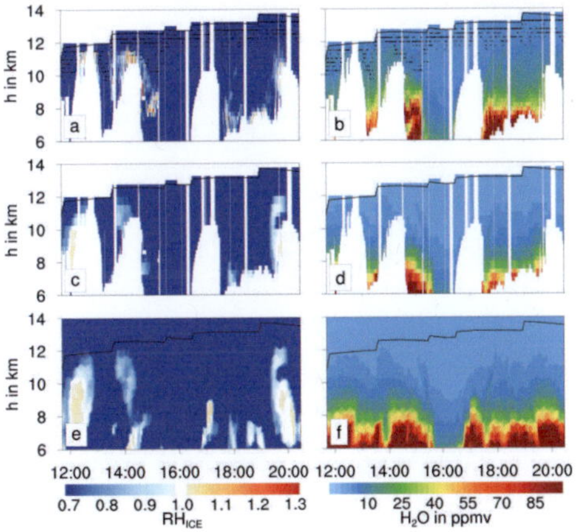

Figure E.5: Cross sections of relative humidity (i.e. supersaturation versus ice phase) (a) and water vapor volume mixing ratio (b) derived from GLORIA measurements during the HALO flight PGS14 on 26 February 2016. Relative humidity (c) and water vapor volume mixing ratio (d) simulated by the model and filtered for geolocations where GLORIA data are available. Full vertical cross section of the same parameters from the model simulation *REF08* (e,f). Times in UTC.

PGS17, 9 March 2016

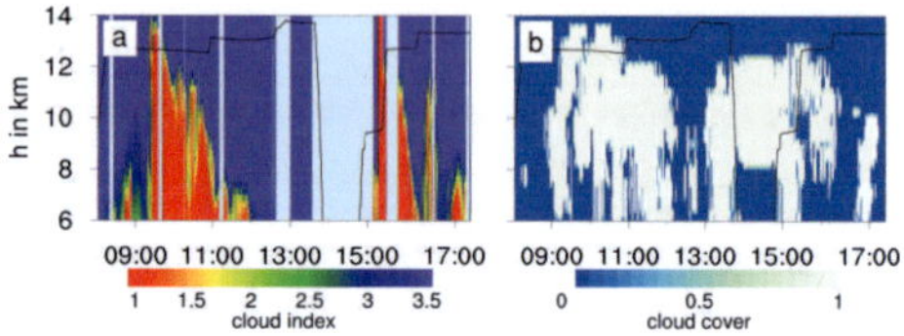

Figure E.6: (a) Vertical cross-section of cloud index derived from GLORIA measurements during PGS17 on 9 March 2016. Low cloud index values indicate opaque conditions due to cloud particles along the instrument's line of sight. (b) Cloud mask derived from the model simulation *REF09* at the geolocations of the measurement (0 = no cloud / 1 = cloud present). Black line in both panels: HALO flight altitude. Times in UTC.

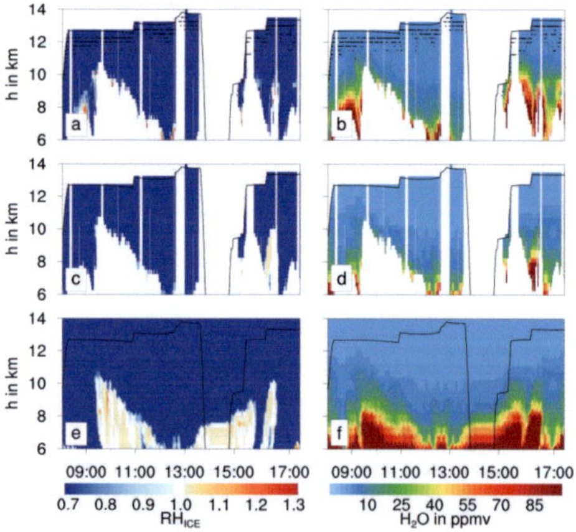

Figure E.7: Cross sections of relative humidity (i.e. supersaturation versus ice phase) (a) and water vapor volume mixing ratio (b) derived from GLORIA measurements during the HALO flight PGS17 on 9 March 2016. Relative humidity (c) and water vapor volume mixing ratio (d) simulated by the model and filtered for geolocations where GLORIA data are available. Full vertical cross section of the same parameters from the model simulation *REF09* (e,f). Times in UTC.

PGS19, 13 March 2016

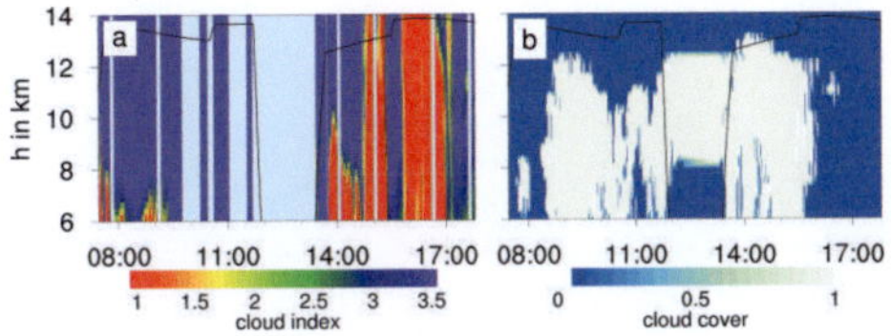

Figure E.8: (a) Vertical cross-section of cloud index derived from GLORIA measurements during PGS19 on 13 March 2016. Low cloud index values indicate opaque conditions due to cloud particles along the instrument's line of sight. (b) Cloud mask derived from the model simulation *REF10* at the geolocations of the measurement (0 = no cloud / 1 = cloud present). Black line in both panels: HALO flight altitude. Times in UTC.

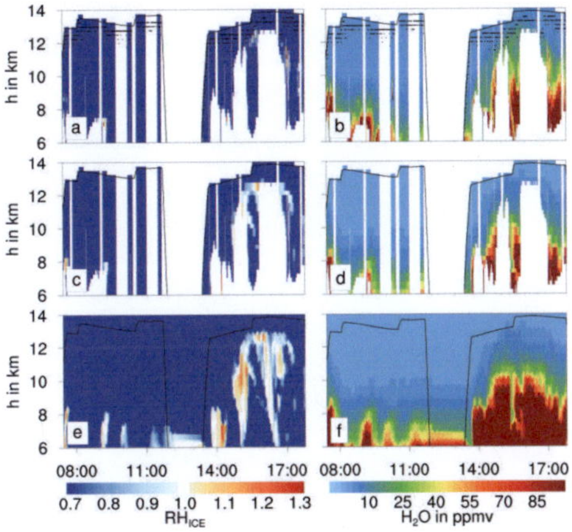

Figure E.9: Cross sections of relative humidity (i.e. supersaturation versus ice phase) (a) and water vapor volume mixing ratio (b) derived from GLORIA measurements during the HALO flight PGS19 on 13 March 2016. Relative humidity (c) and water vapor volume mixing ratio (d) simulated by the model and filtered for geolocations where GLORIA data are available. Full vertical cross section of the same parameters from the model simulation *REF10* (e,f). Times in UTC.

Symbols and Abbreviations

Symbol	Definition	Unit
α_0	space-filling for riming	-
β_{EXT}	extinction coefficient	m^{-1}
Γ	gamma function	-
Δx	effective ICON grid spacing	m
θ	potential temperature or scattering angle	K / °
λ, λ_L	slope parameter of generalized Γ-distribution	-
λ_{Ns}	surviving fraction of ice crystals	-
μ	parameter of generalized Γ-distribution	-
ν	parameter of generalized Γ-distribution	-
ρ	density	$kg\ m^{-3}$
σ_{EXT}	extinction efficiency	-
σ_{SCA}	scattering efficiency	-
σ_i	standard deviation of aerosol mode *i*	-
σ_w	standard deviation of vertical velocity	$m\ s^{-1}$
τ	characteristic time or optical depth	s / -
ω	single scattering albedo	-
A, A_L	scaling parameter of generalized Γ-distribution	$m^{-3}\ kg^{-1}$, m^{-4}

A_c	area of ICON grid cells	m^2
A_r	aspect ratio of hydrometeors	-
a_{geo}	parameter in mass-size relation	-
a_{vel}	parameter in fallspeed relation	-
A_p	statistical average projected area of particle	m^2
b_{geo}	exponent in mass-size relation	-
b_{vel}	exponent in fallspeed relation	-
b_{span}	wing span	m
BiI_3	Bismuth tri-iodide	
c_p	specific heat capacity of air	$J\ kg^{-1}\ K^{-1}$
C	capacity	-
CC	cloud cover	%
CDR	carbon dioxide removal	
CO_2	carbon dioxide	
CTY	cloud type	-
d	flight distance	m
D	size of hydrometeor	m
$d_{k,i}$	median diameter for k^{th} moment of aerosol mode i	μm
d_{HOM}	mean diameter of aerosol droplets for homogeneous nucleation	μm
d_{SEED}	mean diameter for 0th moment of seeding aerosol	μm

D_v	molecular diffusion coefficient	$m^2\ s^{-1}$
ddT_{LW}	longwave radiative cooling rate	$K\ h^{-1}$
E	extinction coefficient	km^{-1}
E_{AB}	collision efficiency for classes A and B	-
EI_{iceno}	emission index for ice crystals	kg^{-1}
f, f_L	particle size distribution	$m^{-3}\ kg^{-1}$, m^{-4}
F_{ven}	ventilation coefficient	-
f_w	factor for TKE-based calculation of σ_w	-
g	asymmetry factor	-
GCM	global circulation model	
h	height	m / km
I	ice mass mixing ratio	$kg\ kg^{-1}$
I_0	water vapor emission per flight distance	$kg\ m^{-1}$
INP	ice nucleating particles	
IWC	ice water content	$mg\ m^{-3}$
J	nucleation rate	$s^{-1}\ m^{-3}$
K_T	conductivity of heat	$J\ m^{-1}\ s^{-1}\ K^{-1}$
L	ice crystal size	m
L_{iv}	latent heat for sublimation	$J\ kg^{-1}$
L_{min}	minimum particle length	m
L_{max}	maximum particle length	m
LES	large-eddy simulation	
LW	longwave	

LWC	liquid water content	mg m^{-3}
LWP	liquid water path	mg m^{-2}
m	mass of hydrometeor	kg
M^{k}	kth moment of $f(x)$	kg^{k} m^{-3}
m_{min}	minimum particle mass	kg
m_{max}	maximum particle mass	kg
M_{x}	molar mass of species x	kg
n	ice crystal number concentration	m^{-3}
N	number concentration	kg^{-1}
N_{BV}	Brunt-Väisälä frequency	s^{-1}
n_{CLD}	number concentration of cloud droplets	l^{-1}
n_{DUST}	number concentration of mineral dust	l^{-1}
n_{HOM}	number concentration of aerosol droplets for homogeneous nucleation	l^{-1}
n_{ICE}	number concentration of ice crystals	l^{-1}
n_{init}	'emitted' ice crystal number concentration	m^{-3}
n_{LIM}	threshold number concentration of INP to inhibit homogeneous nucleation	l^{-1}
n_{SEED}	number concentration of seeding aerosol	l^{-1}
N_{s}	number of surviving ice crystals per flight distance	m^{-1}
N_{0}	number of produced ice crystals per flight distance	m^{-1}

O	area of particle	m^2
OLR	outgoing longwave radiation	$W\ m^{-2}$
p	pressure	hPa
P	scattering phase function	-
p_{sat}	saturation pressure	hPa
PBL	planetary boundary layer	
PSC	polar stratospheric cloud	
PV	photovoltaic	
q_x	mass concentration of species x	$kg\ kg^{-1}$
q_{init}	'emitted' ice crystal mass concentration	$kg\ m^{-3}$
r	radius	μm
r_e	effective radius	μm
r_E	Earth's radius	m
R_d	specific gas constant for dry air	$J\ kg^{-1}\ K^{-1}$
R_v	specific gas constant for water vapor	$J\ kg^{-1}\ K^{-1}$
RANS	Reynolds-Averaged Navier Stokes	
RH_x	relative humidity with respect to phase x	-
S	surface area of particle	m^2
s_{crit}	critical supersaturation for homogeneous nucleation	-
s_i	supersaturation with respect to ice	-
s_{max}	maximum supersaturation for homogeneous nucleation	-

SAI	stratospheric aerosol injection	
SFC	surface	
SRM	solar radiation management	
SW	shortwave	
TKE	turbulent kinetic energy	kg m^2 s^{-2}
t	time	s / UTC
T	temperature	K
TOA	top of atmosphere	
UTLS	upper troposphere / lower stratosphere	
v	velocity	m s^{-1}
V	volume	m^3
w	vertical wind speed	m s^{-1}

Acknowledgment

I acknowledge the use of imagery from the Land Atmosphere Near real-time Capability for EOS (LANCE) system and services from the Global Imagery Browse Services (GIBS), both operated by the NASA/GSFC/Earth Science Data and Information System (ESDIS, `http://earthdata.nasa.gov`) with funding provided by NASA/HQ.
I acknowledge the use of the radio sounding data provided by the University of Wyoming, Department of Atmospheric Science.
I acknowledge the use of CALIPSO data provided by the NASA Langley Research Center's Atmospheric Science Data Center (ASDC).
I would like to acknowledge operational, technical and scientific support during the NSF HIPPO Global campaign provided by NCAR's Earth Observing Laboratory, sponsored by the National Science Foundation.
The PGS campaign was supported by the German Research Foundation (Deutsche Forschungsgemeinschaft, DFG Priority Program SPP 1294). I thank the PGS coordination team, the GLORIA team and DLR-FX for organizing the campaign, performing the GLORIA measurements and realizing the HALO flights.
The source code for the models as well as the data from the model simulations are available from the author upon request.

Finally, it is time to say thank-you to all that have contributed to this dissertation.
At first, I am very grateful to Prof. Christoph Kottmeier for supervising this thesis and for his constant support and mentoring.

Next, I want to acknowledge Prof. Thomas Leisner for co-supervising this thesis, but especially for acquiring the AWiCiT project, for his motivating interest in my work and for a lot of helpful questions and suggestions.
Dr. Bernhard Vogel is the mentor of my academic training since starting with my Bachelor's Thesis. Thank you for a never-ceasing encouragement, for supporting and motivating me with new ideas and your great scientific curiosity.
Also, I am very grateful to Dr. Heike Vogel, for setting me an example of a pragmatic point of view on problems, and for always having some time left to lend assistance and advice with all kind of minor and major issues.
For the nice and constructive atmosphere at work, thanks a lot to my office colleagues Dr. Konrad Deetz, Lukas Muser, Dr. Tobias Schad, and Sven Werchner. Of course, I also express a big thank-you to all the other, both current and former, members of our working group, Dr. Andrew Barrett, Dr. Gholamali Hoshyaripour, Dr. Daniel Rieger, Anika Rohde, Dr. Frank Wagner, and Carolin Walter. Valuable discussions concerning scientific and non-scientific topics, given hints and supported technical problems eased my life a lot.
The ICON-ART development team, especially Jochen Förstner, Dr. Daniel Rieger, Dr. Roland Ruhnke, Dr. Jennifer Schröter, and Michael Weimer, assured themselves my gratitude for the enjoyable collaboration and all kinds of technical support.
For exciting insights into your measurements and modeling approaches related to our project, enlightening questions, and long business trips to remote places, thanks a lot to Steffen Münch, Dr. Isabelle Steinke, and Tobias Schorr.
Also I want to acknowledge Dr. Florian Haenel and Dr. Wolfgang Woiwode for providing me with the GLORIA data, for giving lots of detailed information and explanations, and for devoting lots of time and expertise to working on our manuscript.
I am deeply grateful to Prof. Trude Storelvmo for hosting my visit at University of Oslo. Thanks for lots of time spent discussing, but best of all, for the great

encouraging interest in my work. A substantial part of this thesis originates from this time. Of course, I also want to say a big thank-you to all the other members of the Department of Geosciences for the warm welcome and the friendly atmosphere. I also thank the Graduate School for Climate and Environment (GRACE) for funding my research stay at University of Oslo.

For pleasant collaborations, for providing the revised fits, for double-checking code, and for lots of valuable discussions and advice, I am very grateful to Dr. Ulrich Blahak, Dr. Pavel Khain, Dr. Martin Köhler, and Dr. Harel Muskatel.

Dr. Axel Seifert provided me with lots of explanations and assistance with technical and scientific topics. For this, and for his motivating interest in my work, I want to give my thanks.

My deep gratitude is to be expressed to Dr. Simon Unterstraßer for the fruitful discussions and advices, but most of all, for the endless patience with helping me bringing about my first publication.

Furthermore, I am very grateful to the colleagues in Cologne, Jan Bechtold, Martin Jung and Henry Pak, who provided the flight trajectories and answered all of my questions.

Also, thanks a lot to Gabi Klinck for support with all kind of IT-related issues, for chats at Friday afternoons, and for continuously encouraging me abandoning a bad habit.

Finally, thanks to all those friends and colleagues for having such a nice time alongside and especially besides the PhD. Thanks to my family for their support, patience and understanding throughout the years.

Wissenschaftliche Berichte des Instituts für Meteorologie und Klimaforschung des Karlsruher Instituts für Technologie (0179-5619)

Bisher erschienen:

Nr. 1: *Fiedler, F. / Prenosil, T.*
Das MESOKLIP-Experiment. (Mesoskaliges Klimaprogramm im Oberrheintal). August 1980

Nr. 2: *Tangermann-Dlugi, G.*
Numerische Simulationen atmosphärischer Grenzschichtströmungen über langgestreckten mesoskaligen Hügelketten bei neutraler thermischer Schichtung. August 1982

Nr. 3: *Witte, N.*
Ein numerisches Modell des Wärmehaushalts fließender Gewässer unter Berücksichtigung thermischer Eingriffe. Dezember 1982

Nr. 4: *Fiedler, F. / Höschele, K. (Hrsg.)*
Prof. Dr. Max Diem zum 70. Geburtstag. Februar 1983 (vergriffen)

Nr. 5: *Adrian, G.*
Ein Initialisierungsverfahren für numerische mesoskalige Strömungs modelle. Juli 1985

Nr. 6: *Dorwarth, G.*
Numerische Berechnung des Druckiderstandes typischer Geländeformen. Januar 1986

Nr. 7: *Vogel, B.; Adrian, G. / Fiedler, F.*
MESOKLIP-Analysen der meteorologischen Beobachtungen von mesoskaligen Phänomenen im Oberrheingraben. November 1987

Nr. 8: *Hugelmann, C.-P.*
Differenzenverfahren zur Behandlung der Advektion. Februar 1988

Nr. 9: *Hafner, T.*
Experimentelle Untersuchung zum Druckwiderstand der Alpen. April 1988

Nr. 10: *Corsmeier, U.*
Analyse turbulenter Bewegungsvorgänge in der maritimen atmosphärischen Grenzschicht. Mai 1988

Nr. 11: *Walk, O. / Wieringa, J.(eds)*
Tsumeb Studies of the Tropical Boundary-Layer Climate. Juli 1988

Nr. 12: *Degrazia, G. A.*
Anwendung von Ähnlichkeitsverfahren auf die turbulente Diffusion in der konvektiven und stabilen Grenzschicht. Januar 1989

Nr. 13: *Schädler, G.*
Numerische Simulationen zur Wechselwirkung zwischen Landoberflächen und atmophärischer Grenzschicht. November 1990

Nr. 14: *Heldt, K.*
Untersuchungen zur Überströmung eines mikroskaligen Hindernisses in der Atmosphäre. Juli 1991

Nr. 15: *Vogel, H.*
Verteilungen reaktiver Luftbeimengungen im Lee einer Stadt – Numerische Untersuchungen der relevanten Prozesse. Juli 1991

Nr. 16: *Höschele, K.(ed.)*
Planning Applications of Urban and Building Climatology – Proceedings of the IFHP / CIB-Symposium Berlin, October 14-15, 1991. März 1992

Nr. 17: *Frank, H. P.*
Grenzschichtstruktur in Fronten. März 1992

Nr. 18: *Müller, A.*
Parallelisierung numerischer Verfahren zur Beschreibung von Ausbreitungs- und chemischen Umwandlungsprozessen in der atmosphärischen Grenzschicht. Februar 1996

Nr. 19: *Lenz, C.-J.*
Energieumsetzungen an der Erdoberfläche in gegliedertem Gelände. Juni 1996

Nr. 20: *Schwartz, A.*
Numerische Simulationen zur Massenbilanz chemisch reaktiver Substanzen im mesoskaligen Bereich. November 1996

Nr. 21: *Beheng, K. D.*
Professor Dr. Franz Fiedler zum 60. Geburtstag. Januar 1998

Nr. 22: *Niemann, V.*
Numerische Simulation turbulenter Scherströmungen mit einem Kaskadenmodell. April 1998

Nr. 23: *Koßmann, M.*
Einfluß orographisch induzierter Transportprozesse auf die Struktur der atmosphärischen Grenzschicht und die Verteilung von Spurengasen. April 1998

Nr. 24: *Baldauf, M.*
Die effektive Rauhigkeit über komplexem Gelände – Ein Störungstheoretischer Ansatz. Juni 1998

Nr. 25: *Noppel, H.*
Untersuchung des vertikalen Wärmetransports durch die Hangwindzirkulation auf regionaler Skala. Dezember 1999

Nr. 26: *Kuntze, K.*
Vertikaler Austausch und chemische Umwandlung von Spurenstoffen über topographisch gegliedertem Gelände. Oktober 2001

Nr. 27: *Wilms-Grabe, W.*
Vierdimensionale Datenassimilation als Methode zur Kopplung zweier verschiedenskaliger meteorologischer Modellsysteme. Oktober 2001

Nr. 28: *Grabe, F.*
Simulation der Wechselwirkung zwischen Atmosphäre, Vegetation und Erdoberfläche bei Verwendung unterschiedlicher Parametrisierungsansätze. Januar 2002

Nr. 29: *Riemer, N.*
Numerische Simulationen zur Wirkung des Aerosols auf die troposphärische Chemie und die Sichtweite. Mai 2002

Nr. 30: *Braun, F. J.*
Mesoskalige Modellierung der Bodenhydrologie.
Dezember 2002

Nr. 31: *Kunz, M.*
Simulation von Starkniederschlägen mit langer Andauer über Mittelgebirgen. März 2003

Nr. 32: *Bäumer, D.*
Transport und chemische Umwandlung von Luftschadstoffen im Nahbereich von Autobahnen – numerische Simulationen.
Juni 2003

Nr. 33: *Barthlott, C.*
Kohärente Wirbelstrukturen in der atmosphärischen Grenzschicht. Juni 2003

Nr. 34: *Wieser, A.*
Messung turbulenter Spurengasflüsse vom Flugzeug aus.
Januar 2005

Nr. 35: *Blahak, U.*
Analyse des Extinktionseffektes bei Niederschlagsmessungen mit einem C-Band Radar anhand von Simulation und Messung.
Februar 2005

Nr. 36: *Bertram, I.*
Bestimmung der Wasser- und Eismasse hochreichender konvektiver Wolken anhand von Radardaten, Modellergebnissen und konzeptioneller Betrachtungen. Mai 2005

Nr. 37: *Schmoeckel, J.*
Orographischer Einfluss auf die Strömung abgeleitet aus Sturmschäden im Schwarzwald während des Orkans „Lothar". Mai 2006

Nr. 38: *Schmitt, C.*
Interannual Variability in Antarctic Sea Ice Motion: Interannuelle Variabilität antarktischer Meereis-Drift.
Mai 2006

Nr. 39: *Hasel, M.*
Strukturmerkmale und Modelldarstellung der Konvektion über Mittelgebirgen. Juli 2006

Ab Band 40 erscheinen die Wissenschaftlichen Berichte des Instituts für Meteorologie und Klimaforschung bei KIT Scientific Publishing (ISSN 0179-5619). Die Bände sind unter www.ksp.kit.edu als PDF frei verfügbar oder als Druckausgabe bestellbar.

Nr. 40: *Lux, R.*
Modellsimulationen zur Strömungsverstärkung von orographischen Grundstrukturen bei Sturmsituationen
ISBN 978-3-86644-140-8

Nr. 41: *Straub, W.*
Der Einfluss von Gebirgswellen auf die Initiierung und Entwicklung konvektiver Wolken
ISBN 978-3-86644-226-9

Nr. 42: *Meißner, C.*
High-resolution sensitivity studies with the regional climate model COSMO-CLM
ISBN 978-3-86644-228-3

Nr. 43: *Höpfner, M.*
Charakterisierung polarer stratosphärischer Wolken mittels hochauflösender Infrarotspektroskopie
ISBN 978-3-86644-294-8

Nr. 44: *Rings, J.*
Monitoring the water content evolution of dikes
ISBN 978-3-86644-321-1

Nr. 45: *Riemer, M.*
Außertropische Umwandlung tropischer Wirbelstürme: Einfluss auf das Strömungsmuster in den mittleren Breiten
ISBN 978-3-86644-766-0

Nr. 46: Nicht erschienen.

Nr. 47: Nicht erschienen.

Nr. 48: Nicht erschienen.

Nr. 49: *Peters, T.*
Ableitung einer Beziehung zwischen der Radarreflektivität, der Niederschlagsrate und weiteren aus Radardaten abgeleiteten Parametern unter Verwendung von Methoden der multivariaten Statistik
ISBN 978-3-86644-323-5

Nr. 50: *Khodayar Pardo, S.*
High-resolution analysis of the initiation of deep convection forced by boundary-layer processes
ISBN 978-3-86644-770-7

Nr. 51: *Träumner, K.*
Einmischprozesse am Oberrand der konvektiven atmosphärischen Grenzschicht
ISBN 978-3-86644-771-4

Nr. 52: Nicht erschienen.

Nr. 53: *Lundgren, K.*
Direct Radiative Effects of Sea Salt on the Regional Scale
ISBN 978-3-86644-773-8

Nr. 54: *Sasse, R.*
Analyse des regionalen atmosphärischen Wasserhaushalts unter Verwendung von COSMO-Simulationen und GPS-Beobachtungen
ISBN 978-3-86644-774-5

Nr. 55: *Grenzhäuser, J.*
Entwicklung neuartiger Mess- und Auswertungs-strategien für ein scannendes Wolkenradar und deren Anwendungsbereiche
ISBN 978-3-86644-775-2

Nr. 56: *Grams, C.*
Quantification of the downstream impact of extratropical transition for Typhoon Jangmi and other case studies
ISBN 978-3-86644-776-9

Nr. 57: *Keller, J.*
Diagnosing the Downstream Impact of Extratropical Transition Using Multimodel Operational Ensemble Prediction Systems
ISBN 978-3-86644-984-8

Nr. 58: *Mohr, S.*
Änderung des Gewitter- und Hagelpotentials im Klimawandel
ISBN 978-3-86644-994-7

Nr. 59: *Puskeiler, M.*
Radarbasierte Analyse der Hagelgefährdung in Deutschland
ISBN 978-3-7315-0028-5

Nr. 60: *Zeng, Y.*
Efficient Radar Forward Operator for Operational
Data Assimilation within the COSMO-model
ISBN 978-3-7315-0128-2

Nr. 61: Nicht erschienen.

Nr. 62: *Jerger, D.*
Radar Forward Operator for Verification of Cloud
Resolving Simulations within the COSMO Model
ISBN 978-3-7315-0172-5

Nr. 63: *Maurer, V.*
Vorhersagbarkeit konvektiver Niederschläge :
Hochauflösende Ensemblesimulationen für Westafrika
ISBN 978-3-7315-0189-3

Nr. 64: *Stawiarski, C.*
Optimizing Dual-Doppler Lidar Measurements of Surface
Layer Coherent Structures with Large-Eddy Simulations
ISBN 978-3-7315-0197-8

Nr. 65: *Mahlke, H.*
Mechanismen der Auslösung hochreichender Konvektion
im südwestdeutschen Mittelgebirgsraum
ISBN 978-3-7315-0203-6

Nr. 66: *Fosser, G.*
Precipitation statistics from regional climate model
at resolutions relevant for soil erosion
ISBN 978-3-7315-0227-2

Nr. 67: *Adler, B.*
Boundary-Layer Processes Producing Mesoscale
Water-Vapour Variability over a Mountainous Island
ISBN 978-3-7315-0247-0

Die Bände sind unter www.ksp.kit.edu als PDF frei verfügbar oder als Druckausgabe bestellbar.

Nr. 68: *Kraut, I.*
Separating the Aerosol Effect in Case of a „Medicane"
ISBN 978-3-7315-0405-4

Nr. 69: *Breil, M.*
Einfluss der Boden-Vegetation-Atmosphären Wechselwirkungen auf die dekadische Vorhersagbarkeit des Westafrikanischen Monsuns
ISBN 978-3-7315-0420-7

Nr. 70: *Lott, F. F.*
Wind Systems in the Dead Sea and Footprints in Seismic Records
ISBN 978-3-7315-0596-9

Nr. 71: *Rieger, D.*
Der Einfluss von natürlichem Aerosol auf Wolken über Mitteleuropa
ISBN 978-3-7315-0672-0

Nr. 72: *Loewe, K.*
Arctic mixed-phase clouds. Macro- and microphysical insights with a numerical model
ISBN 978-3-7315-0686-7

Nr. 73: *Piper, D. A.*
Untersuchung der Gewitteraktivität und der relevanten großräumigen Steuerungsmechanismen über Mittel- und Westeuropa
ISBN 978-3-7315-0701-7

Nr. 74: *Metzger, J.*
Wind Systems and Energy Balance in the Dead Sea Valley
ISBN 978-3-7315-0699-7

Nr. 75: *Deetz, K.*
Assessing the Aerosol Impact on Southern West African Clouds and Atmospheric Dynamics
ISBN 978-3-7315-0744-4

Nr. 76: *Ehmele, F. T.*
Stochastische Simulation großflächiger, hochwasserrelevanter Niederschlagsereignisse.
ISBN 978-3-7315-0761-1

Nr. 77: *Hackenbruch, J.*
Anpassungsrelevante Klimaänderungen für städtische Baustrukturen und Wohnquartiere
ISBN 978-3-7315-0771-0

Nr. 78: *Schmidberger, M.*
Hagelgefährdung und Hagelrisiko in Deutschland basierend auf einer Kombination von Radardaten und Versicherungsdaten.
ISBN 978-3-7315-0846-5

Nr. 79: *Gruber, S.*
Contrails and Climate Engineering - Process Studies on Natural and Artificial High-Level Clouds and Their Impact on the Radiative Fluxes.
ISBN 978-3-7315-0896-0

Die Bände sind unter www.ksp.kit.edu als PDF frei verfügbar oder als Druckausgabe bestellbar.